**Editor**
Lorin Klistoff, M.A.

**Managing Editor**
Karen J. Goldfluss, M.S. Ed.

**Illustrator**
Kelly McMahon

**Cover Artist**
Brenda DiAntonis

**Creative Director**
Karen J. Goldfluss, M.S. Ed.

**Art Coordinator**
Renée Christine Yates

**Art Production Manager**
Kevin Barnes

**Imaging**
Denise Thomas
Nathan Rivera

*Publisher*

*Mary D. Smith, M.S. Ed.*

**Authors**

# Nicole Bauer and Judy Tertini

(Revised and rewritten by Teacher Created Resources, Inc.)

*Teacher Created Resources, Inc.*
6421 Industry Way
Westminster, CA 92683
www.teachercreated.com
**ISBN-13: 978-1-4206-8988-4**

© *2007 Teacher Created Resources, Inc.*

Made in U.S.A.

# Table of Contents

# Table of Contents

# Introduction

## Targeting Math

The series *Targeting Math* is a comprehensive classroom resource. It has been developed so that teachers can find activities and reproducible pages for all areas of the math curriculum.

## About This Series

The twelve books in this series cover all aspects of the math curriculum in an easy-to-access format. Each of the three levels has four books: Numeration and Fractions; Operations and Number Patterns; Geometry, Chance and Data; and Measurement. Each topic in a book is covered by one or more units that are progressive in level. The teacher is able to find resources for all students whatever their ability. This enables the teacher to differentiate for different ability groups. It also provides an easy way to find worksheets at different levels for remediation and extension.

## About This Book

*Targeting Math: Measurement (Grades 1 & 2)* contains the following topics: area, capacity/volume, length, mass, and time. Every topic contains at least two complete units of work. (See Table of Contents for specific skills.)

## About Each Unit

Each unit is complete in itself. It begins with a list of objectives, resources needed, mathematical language used, and a description of each student activity page. This is followed by suggested student activities to reinforce learning. The reproducible pages cover different aspects of the topic in a progressive nature and all answers are included. Every unit includes an assessment page. These assessment pages are important resources in themselves as teachers can use them to find out what their students know about a new topic. They can also be used for assessing specific outcomes when clear feedback is needed.

## About the Skills Index

A Skills Index is provided at the end of the book. It lists specific objectives for the student pages of each unit in the book.

# AREA

These units provide opportunities to estimate, measure, order, and compare areas using formal and informal units of measurement. Greater, smaller, and equal areas are modeled. Students are involved in decision making and lateral-thinking activities. Grid paper and squares are used to calculate areas. Informal exploration of surface area is carried out on the box shapes of a cube and rectangular prism. There are two assessment pages and an activity, which involves identifying the largest and smallest areas.

© Teacher Created Resources, Inc.

#8988 Targeting Math: Measurement

# BEGINNING AREA

## Unit 1

Comparing areas
Greater area
Estimating
Ordering area
Measuring

## Objectives

- estimate, compare, and order the areas of shapes using informal units
- recognize and compare the sizes of groups through a variety of strategies such as estimating, matching one-to-one, and counting
- make non-numerical estimates of size involving everyday movements and actions
- respond to and use everyday comparative language relating to area when recording and communicating measurements
- sort and describe objects in terms of their features such as size and shape
- explain simple mathematical situations using everyday language, actions, materials, and drawings

## Language

estimate, how many, greater area, closed shapes, same size, order, smallest area, biggest area, medium-sized, squares, count, actual, grid

## Materials/Resources

colored pencils, scissors, glue, grid paper, plastic tiles, counters

## Contents of Student Pages

* Materials needed for each reproducible student page

### Page 8 Informal Measuring
cutting out, estimating, and then pasting ladybugs on a leaf

* colored pencils, scissors, glue

### Page 9 Greater Area
finding closed shapes; circling objects with greater area; drawing two objects and circling the one with greater area

* colored pencils

### Page 10 More Greater Area
drawing items as instructed

* colored pencils

### Page 11 Ordering Areas
finding shapes that are the same size; ordering shapes according to size

### Page 12 Ordering More Areas
ordering three paddocks according to area; cutting out animals and pasting in specified paddock

* colored pencils, scissors, glue

### Page 13 Finding Areas
finding areas by counting squares; drawing shapes on grid paper and finding areas by counting squares; finding area of shapes by using informal units; recording findings

* grid paper, counters

### Page 14 Assessment
* colored pencils

### Page 15 Activity—Largest and Smallest
finding the largest and smallest areas

* colored pencils

## Remember

- ❑ Give students many opportunities to find, compare, and order areas of a variety of objects using a variety of informal measures.
- ❑ Encourage students to estimate and then check.

6

# Additional Activities

- Students trace around their hands, compare the areas, and make a chart. Get cutouts of other people's hands (Examples: a baby or a person with large hands). Label the hands that have the greatest and smallest areas. Do the same for feet.
- Look in reference books and find out the size of dinosaurs' feet. Encourage students to make cardboard cutouts.
- Trace around two students. Students cover the areas using one type of flat object. Compare the areas. Display.
- If you have a home center, provide different-sized sheets and tablecloths. Discuss sizes.
- Provide puzzles, mosaics, tangrams, blocks, paintings, and collages. All of these provide opportunities for students to cover areas.
- Collect boxes of different sizes and used wrapping paper. Students estimate how much paper they will need to wrap the boxes.
- Go for a walk around the neighborhood. Students observe areas such as driveways.
- Have a "feely box." Put in two objects that have different areas. The student identifies the greater or smaller area.
- Make pancakes with the students. Make the pancakes so there is a variety of sizes. Discuss the areas. Draw around these. Use different informal units to measure the areas with the students.
- Draw different shapes in the playground sand or dirt. Students estimate how many students are needed to cover the areas.
- Bring in towels of different sizes. Students order according to area. Check by putting the towels on top of each other.
- Take opportunities in the classroom to draw students' attention to area comparison (Example: areas of different-sized paper). Compare sizes. Measure using informal units, discuss results, and display some samples. Compare other items (Examples: envelopes, books, stamps).
- If the school has a doll house that needs renovating, revamp it with the students' help. Measuring wallpaper/linoleum/carpet provides excellent opportunities. Does it fit? Is it too big/too small?
- Obtain a large cardboard box and allow the students to wallpaper it and paint the walls. Make windows and a door in it. When finished, students can play in it or have it as a quiet "reading area."

# Answers

## Page 8   Informal Measuring
1. Check to make sure estimate is reasonable.
2. Check to make sure the ladybugs and leaf are colored.
3. 12 ladybugs (all ladybugs should fit on leaf)

## Page 9   Greater Area
1. Check to make sure the five closed shapes are colored.
2. Check to make sure the object with the bigger area is circled in each section.
3. Check to make sure two objects are drawn and the object with the greater area is circled.

## Page 10   More Greater Area
1. Three fish should be drawn in the bigger pond.
2. A design should be drawn on the bigger butterfly.
3. Four children should be drawn in the bigger pool.
4. Flowers should be drawn in the larger garden.
5. Two children should be drawn on the bigger rug.

## Page 11   Ordering Areas
1. a. Third shape should be circled.
   b. Third shape should be circled.
   c. Last shape should be circled.
2. a. 4, 2, 1, 3
   b. 3, 1, 2, 4
   c. 2, 4, 3, 1

## Page 12   Ordering More Areas
1. Make sure paddocks are colored.
2. Make sure animals are colored and cut out.
3. Make sure 6 sheep are in the biggest paddock, 1 sheep is in the smallest paddock, 2 sheep are in the medium-sized paddock, and 4 chickens are in the biggest paddock.
4. a. 10 animals
   b. 1 animal

## Page 13   Finding Areas
1. a. 15     c. 8     e. 10     g. 13
   b. 8      d. 13    f. 10
2. Check to make sure the number of squares matches each shape.
3. Check to make sure there is a reasonable estimate and actual answer.

## Page 14   Assessment
1. Shapes 2 and 3 are colored.
2. Check to make sure the bigger mat in each box is circled.
3. The first butterfly is circled.
4. the third shape
5. 5, 4, 1, 2, 3
6. Check to make sure 4 ducks are drawn in the bigger pond.

## Page 15   Activity—Largest to Smallest
Check to make sure the largest area in each shape is colored red and the smallest area is green.

| **Name** | **Date** |
|---|---|

1. How many ladybugs fit on the leaf?  Estimate

2. Color the ladybugs and the leaf.

3. Cut out the ladybugs to see how many will fit on the leaf.

4. Glue the ladybugs to the leaf.

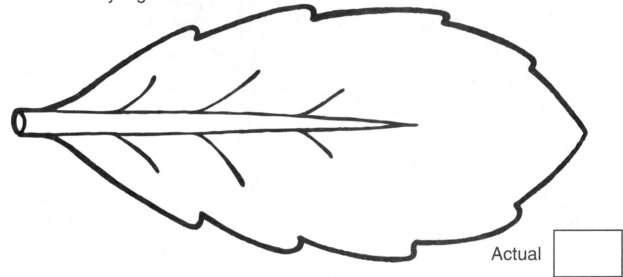

Actual

 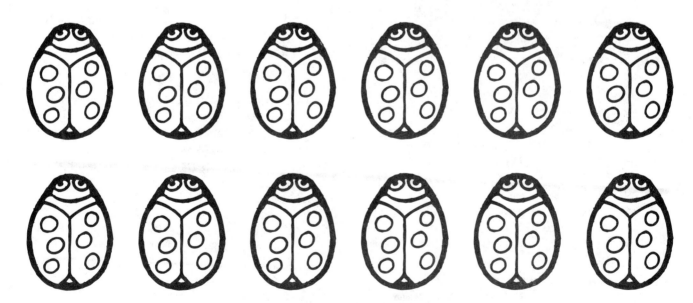

8

**Name**                             **Date**

**1.** Color the closed shapes.

**2.** Circle the object in each pair that has the bigger area.

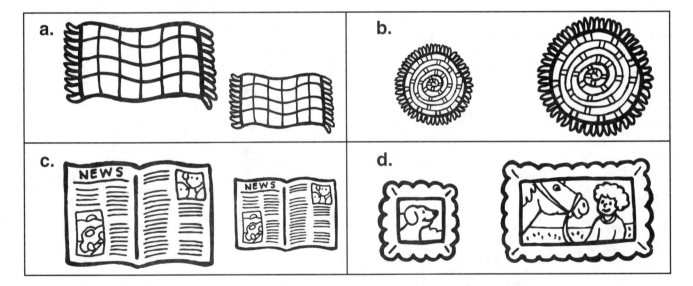

**3.** Draw two objects (one larger than the other). Circle the one that has the greater area.

              *#8988 Targeting Math: Measurement*

**Name**                                    **Date**

1. Draw three fish in the pond with the greater area.

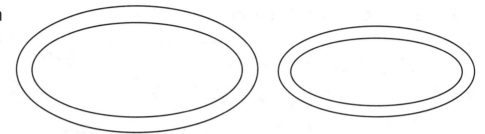

2. Draw a design on the butterfly with the greater area.

3. Draw four children in the pool with the greater area.

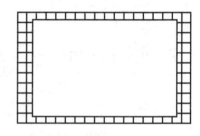

4. Draw flowers in the garden with the greater area.

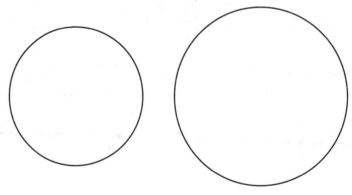

5. Draw two children having a picnic on the rug with the greater area.

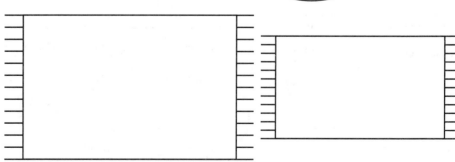

10

| **Name** | **Date** |
|---|---|

1. Circle the shape that is the same size.

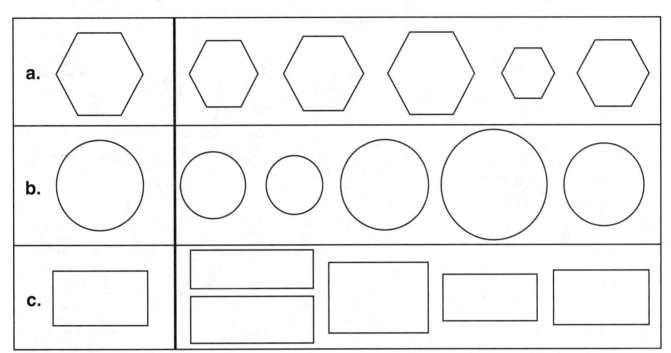

2. Number the shapes from the smallest area (1) to the biggest area (4).

a.

_____    _____    _____    _____

b.

_____    _____    _____    _____

c.

_____    _____    _____    _____

*#8988 Targeting Math: Measurement*

**Name**                      **Date**

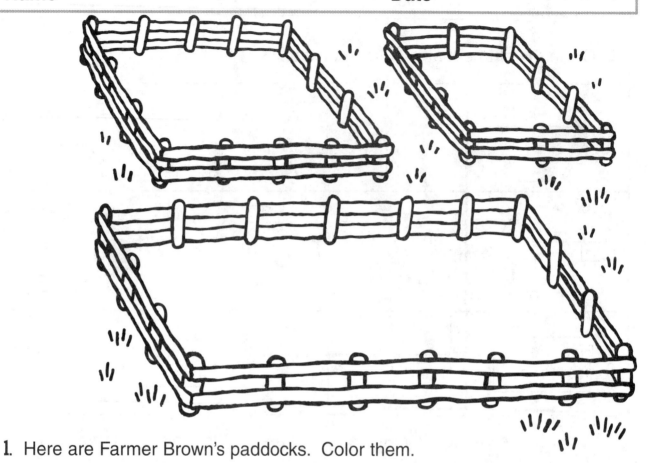

1. Here are Farmer Brown's paddocks. Color them.

2. Color the animals below, then cut them out.

3. Glue 6 sheep in the biggest paddock, 1 sheep in the smallest paddock, and 2 sheep in the medium-sized paddock. Glue 4 chickens in the biggest paddock.

4.  **a.** How many animals are in the biggest paddock? _____

    **b.** How many animals are in the smallest paddock? _____

| **Name** | **Date** |
|---|---|

1. Find the areas of these shapes by counting the squares.

**a.** 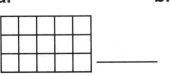 _____

**b.** 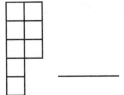 _____

**c.** 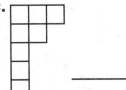 _____

**d.** 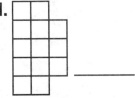 _____

**e.** 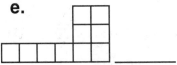 _____

**f.** 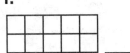 _____

**g.** 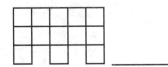 _____

2. Draw some shapes on grid paper. Find the areas by counting the squares. Write the number of squares on each shape.

3. Find the areas of these shapes using plastic counters. Estimate first. Then use the counters to find the areas.

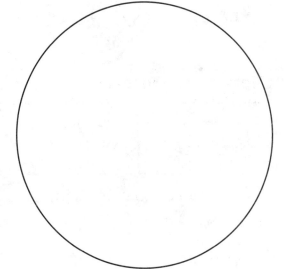

**a.** Estimate: _____

Actual: _____

What did you find? _____

_____

_____

**b.** Estimate: _____

Actual: _____

What did you find? _____

_____

_____

13

| **Name** | **Date** |
|---|---|

1. Color the closed shapes.

2. Circle the mat in each pair that has the greater area.

a.   b.   c.

3. Circle the butterfly that has the greater area.

4. Circle the shape that is the same size.

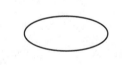  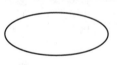 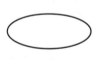 

5. Number the shapes from the smallest to the biggest area.

___  ___  ___  ___  ___

6. Draw 4 ducks on the pond with the greater area.

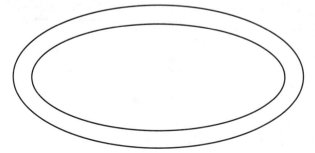 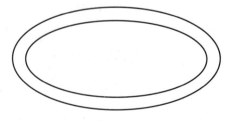

14

**Name**

**Date**

1.

2.

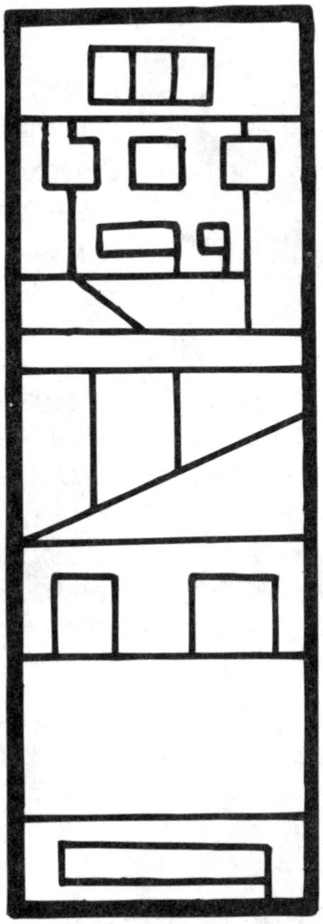

3.

4.

In each shape color:

    **a.** the largest area red.

    **b.** the smallest area green.

Use blue, yellow, and orange to color the rest to make a picture.

15

# MORE AREA

## Unit 2

*Area*
*Measuring*
*Estimating*
*Ordering area*
*Counting*
*Nets*

## Objectives

- *estimate, compare, and order the areas of shapes using informal units*
- *estimate, compare, and order the areas of shapes using formal units*
- *estimate the order of things by length, area, mass, and capacity and make numerical estimates of length using a unit that can be seen or handled*

## Language

*area, counting, number, largest area, smallest area, count, shapes, different shapes, greatest area, estimate, actual, how many more, grid, same area as, cubes, same-sized*

## Materials/Resources

*1 cm grid paper, joining cubes, colored pencils, sticky tape, scissors*

## Contents of Student Page

* *Materials needed for each reproducible student page*

### Page 18  Finding Area

*comparing area by counting squares on a grid; drawing around a hand and foot on grid paper and finding areas*

* *grid paper*

### Page 19  Same Area

*modeling shapes and the identifying shapes of same area; making shapes out of joining cubes*

* *joining cubes*

### Page 20  Estimate Area

*measuring and estimating shapes; identifying greater, smaller, same areas*

### Page 21  Area of Rooms

*Measuring and estimating areas*

### Page 22  Area of Boxes

*measuring areas of boxes; making largest area into a box (optional)*

* *colored pencils, sticky tape (optional), scissors (optional)*

### Page 23  Assessment

* *red and yellow colored pencils*

## Remember

- ❑ *Provide many opportunities for students to estimate areas and then check using a variety of informal units.*
- ❑ *Reinforce the concept that perimeter is different than area. If students are having difficulty, provide more experiences.*

## Additional Activities

❑ *With the students, draw around shadows in the playground. Find the areas using informal units. Compare the areas.*

❑ *Provide geoboards for the students to make different shapes and find the areas of the shapes. These can be drawn on dotted paper and labeled with the size of the area. What is the smallest/largest area that can be found?*

❑ *Encourage students to find objects that have different areas (Examples: plates, leaves). Make cut-outs and compare the areas by covering them with plastic counters.*

❑ *Look up butterflies in the Guiness Book of Records. Students draw the largest wingspan and find its area.*

❑ *Follow up by encouraging students to suggest books to trace around on grid paper to find their areas. Record the results. Label the outlines. Which book cover has the largest/smallest area?*

❑ *Follow up by providing boxes for students to open out and find their areas using informal units.*

❑ *Give students opportunities to develop strategies for counting grid units so that no units are missed or counted twice. Brainstorm what they could do when confronted with half squares, etc.*

❑ *In pairs, students draw identical shapes on 1 cm grid paper. They take turns to throw a die. With each throw, they cover the number of squares with centicubes. The object of the game is to be the first to cover their shape. Alternatively, the squares could be colored in.*

## Answers

### Page 18  Finding Area

1. Phone Book 30
   Tiny Tales 4
   Mouse Book 20
   Big Book 42
   Tall Tales 10
   Atlas 30
2. a. Big Book
   b. Tiny Tales
   c. Phone Book and Atlas
3. Check to make sure the number of squares is correct.

### Page 19  Same Area

1. a. 3, second shape
   b. 5, first shape
   c. 8, second shape
   d. 6, third shape
2. Teacher to check.

### Page 20  Estimate Area

Make sure all estimates are reasonable. Below indicates the actual answer.

1. a. 17
   b. 18
   c. 24
   d. 32
   e. 24
   f. 34
2. a. The letter "g" should be written on the shape in 1. (f.).
   b. The letter "s" should be written on the shape in 1. (a.).
   c. c and e
   d. 24 – 18 = 6

### Page 21  Area of Rooms

1. a. lounge room
   b. Alice's room
   c. Make sure all estimates are reasonable.
      hall = 16 squares
      dining room = 19 squares
      main bedroom = 20 squares
      lounge room = 24 squares
      John's room = 12 squares
      Alice's room = 9 squares
2. a. kitchen
   b. bathroom
   c. kitchen = 8 squares
      bathroom = 7 squares

### Page 22  Area of Boxes

1. a. 70 squares
   b. 30 squares
   c. 90 squares
2. The shape in letter "c" should be decorated.

### Page 23  Assessment

1. a. 10 squares
   b. 15 squares
   c. 8 squares
      Rectangle "c" should be colored red, and rectangle "b" should be colored yellow.
2. 6 squares, The last shape should be circled.
3. Make sure all estimates are reasonable.
   a. 17 squares
   b. 16 squares
   c. 20 squares
4. a. The last shape should have a "g" on it.
   b. The middle shape should have a "s" on it.

| **Name** | **Date** |
|---|---|

1. Find the area of each book by counting the number of squares it covers.

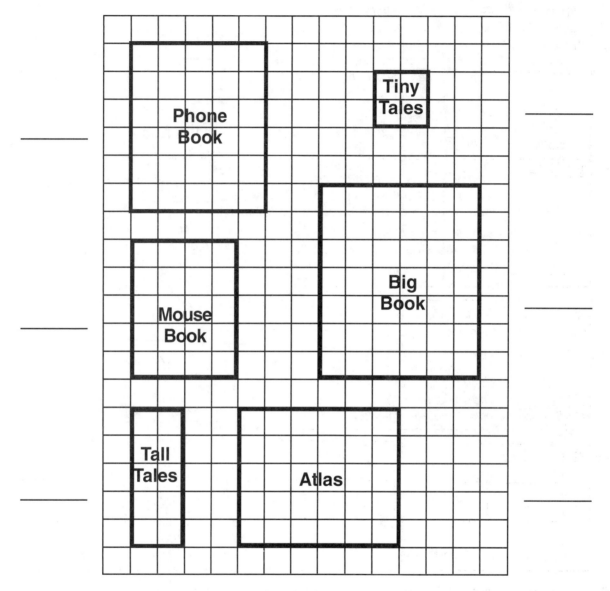

2. **a.** Which book covers the largest area?_____

   **b.** Which book covers the smallest area? _____

   **c.** Which two books cover the same area? _____

3. **a.** On grid paper, draw around your hand. Count the squares it covers and write the number on the hand.

   **b.** On grid paper, draw around your shoe. Count the squares it covers and write the number on the shoe.

**18**

| **Name** | **Date** |
| --- | --- |

1. Record the area of the shape in the box. Then circle the shape that has the same area as the one in the box.

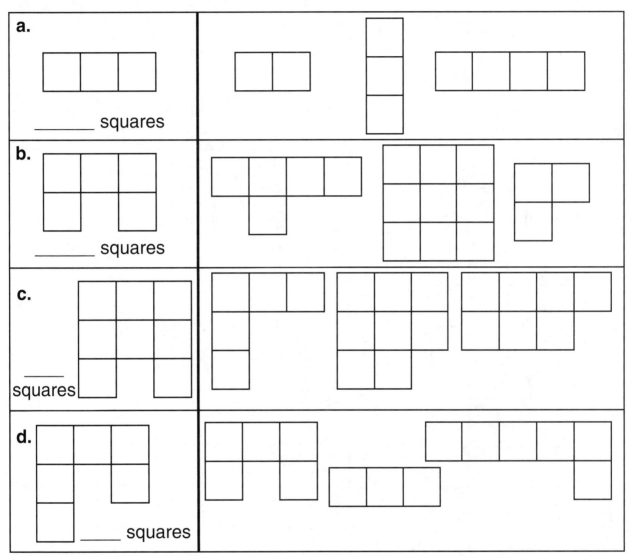

**a.**

_____ squares

**b.**

_____ squares

**c.**

____ squares

**d.**

_____ squares

2. Trace around four joining cubes. Make as many different shapes as you can that have an area of four cubes.

| Name | Date |
|------|------|

1. Find the areas of these shapes. Estimate first.

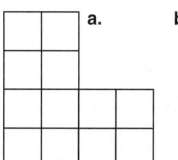

**a.** 　　**b.**

Estimate: ____

Actual: ____

**c.**

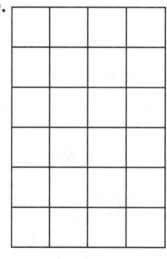

Estimate: ____　　Actual: ____

Estimate: ____　　Actual: ____

**d.**　　**e.**　　**f.**

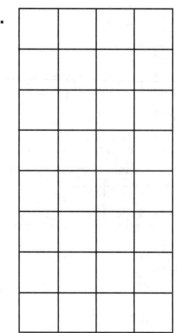

Estimate: ____　　Actual: ____

Estimate: ____

Actual: ____

Estimate: ____

Actual: ____

2. **a.** Which shape has the greatest area? Write **g** on it.

   **b.** Which shape has the smallest area? Write **s** on it.

   **c.** Which two shapes have the same area? _____

   **d.** How many more squares has **c** than **b**? Write a number sentence.

   _____

#8988 *Targeting Math: Measurement*　　　　　　　　　© *Teacher Created Resources, Inc.*

| **Name** | **Date** |
|---|---|

1. The carpet layer has to carpet these rooms.

    **a.** Which room do you think has the greatest area? _____

    **b.** Which room do you think has the smallest area? _____

    **c.** Estimate the number of squares first.  Then, write the actual answer.

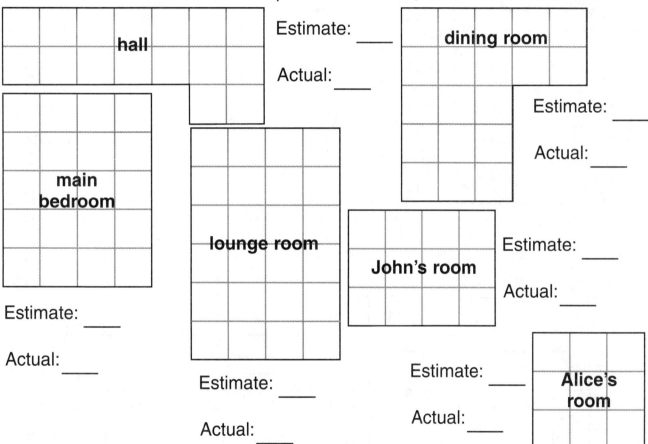

2. The tile person has to tile these rooms.

    **a.** Which room do you think has the greatest tiled area? _____

    **b.** Which room do you think has the smallest tiled area? _____

    **c.** Estimate the number of squares first.  Then, write the actual answer.

*HINT:  Do not count tile under benches, the bath, or sinks.*

㉑

| **Name** | **Date** |
|---|---|

1. Ian opened up three boxes and found these shapes. Find the area of each shape.

a.

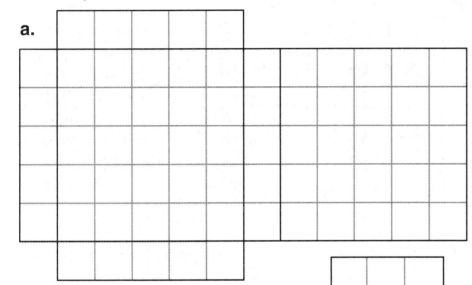

Area = _____

b.

Area = _____

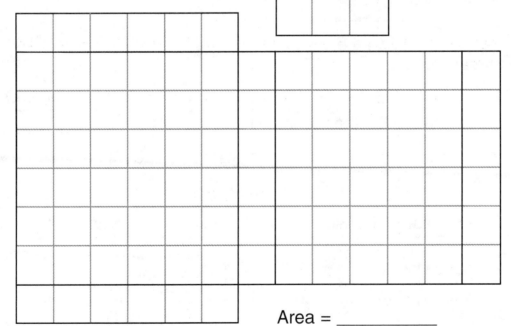

c.

Area = _____

2. Decorate the shape with the largest area. (*Option:* Cut out the shape and make it into a box.)

(22)

| **Name** | **Date** |
|---|---|

**1.** Measure the area of these rectangles by counting the number of squares they cover.

**a.** **b.** _____ **c.** _____

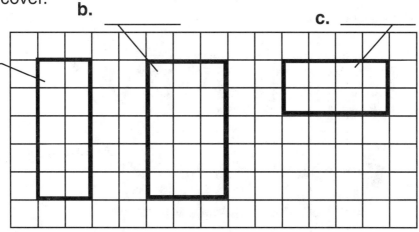

Color the rectangle with the smallest area red and the rectangle with the greatest area yellow.

**2.** Record the area of the first shape. Circle the shape that has the same area.

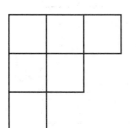 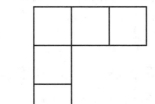 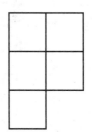 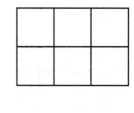

**3.** Draw the same-sized squares as above to find the area of the shapes below. Estimate first.

**a.**

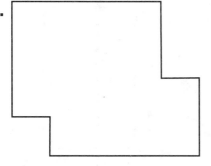

**b.**

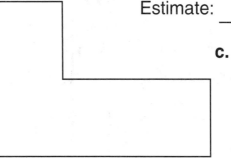

Estimate: ____ Actual: ____

**c.**

Estimate: ____ Actual: ____ Estimate: ____ Actual: ____

**4.** **a.** Which shape has the greatest area? Write **g** on it.

   **b.** Which shape has the smallest area? Write **s** on it.

(23)

# CAPACITY/ VOLUME

These units provide many activities that give students the opportunity to experience capacity and volume in a hands-on way. Students measure with a variety of informal units making use of everyday kitchen containers. The skills of estimating, comparing, and graphing are encouraged. Recognizing full and empty, graphs using capacity, ordering quantities, and problem solving all help to consolidate the concepts covered. There are two assessment pages. The activity page is an experiment using containers and a 50 mL measure.

# BEGINNING CAPACITY/VOLUME

## Unit 1

**Empty, full**
**Identifying capacity**
**Subtraction**
**Ordering**
**Problem solving**

## Objectives

- sort and describe objects in terms of their features such as size and shape
- ask and respond to mathematical questions drawing, describing, acting, telling, guessing and checking, and retelling
- represent addition and subtraction facts up to 20 using concrete materials and in symbolic form
- estimate, compare, and order the capacity of containers using informal units
- describe and model the relationships between the parts and the whole

## Language

full, empty, hold the most, how many more, can take, already, matrix, half-full, order, smallest, largest, half, number sentence, how much left, altogether

## Materials/Resources

colored pencils, (optional: water, glasses, jars)

## Contents of Student Pages

* Materials needed for each reproducible student page

### Page 27 Full and Empty
circling the items that are empty; tracing over words full and empty

### Page 28 Capacity of Containers
identifying containers that would hold the most
* colored pencils

### Page 29 Problem Solving
using subtraction to find how many more

### Page 30 Matrix
completing and labeling the matrix

### Page 31 Order
identifying full and empty; drawing more; ordering items from smallest to largest capacity
* colored pencils

### Page 32 More Problem Solving
solving problems by drawing
* colored pencils, (optional: water, jars, glasses)

### Page 33 Assessment
### Page 34 Activity—Capacity Experiment
* water, jars

## Remember

❏ Sand and water form a necessary part in volume/capacity concept development and in problem solving—some should be available for use.

❏ Estimation should be encouraged.

❏ Capacity is the amount a container can hold. Volume is the amount of space occupied by an object.

25

## Additional Activities

❑ *If possible, provide a water tray and lots of containers in different shapes and sizes for students to explore capacity/volume. Encourage language such as "full" and "empty." Encourage students to estimate before finding out how many of one container are needed to fill another container.*

❑ *Provide substances like sand, rice, or large seeds and a variety of containers in different shapes and sizes for students to explore volume/ capacity.*

❑ *Read Pamela Allen's <u>Mr. Archimedes' Bath</u>, Henry Pluckrose's <u>Capacity</u>, and the story <u>Tiddalick</u> to the students.*

❑ *Allow the students to make a class shop and provide shopping bags for student to pack and unpack items. Encourage language such as "full" and "empty."*

❑ *Have discussions (and picture talks, if you can find suitable illustrations) with students to relate filling and packing activities to things with which they are familiar (Examples: filling the gas tank in a car or filling a bath).*

❑ *Provide nesting toys for the students to manipulate.*

❑ *Encourage the students to work in small groups and see which group can fit the most items into a box.*

❑ *Give students the opportunity to develop language such as the following: "half full," "half empty," and "over full."*

❑ *Provide opportunities to identify substances that are easily poured or emptied, and to identify appropriate containers to hold particular substances.*

❑ *Provide a variety of boxes, glue, tape, etc., for students to make structures.*

## Answers

### Page 27  Full and Empty
1. Check to make sure the words are traced over.
2. The following items should be circled: the empty yogurt container, the empty egg holder, the empty bowl, the empty pail, the empty tub, the empty stroller, the empty toy box, the empty pitcher, the empty cup, and the empty wheelbarrow.

### Page 28  Capacity of Containers
Check to make sure the following are colored:

1. the larger jar
2. the larger pail
3. the larger jug
4. the larger bottle
5. the larger flowerpot
6. the larger box
7. the larger trash can
8. the larger cup

### Page 29  Problem Solving
1. 7
2. 20 – 14 =, 6 (Make sure 6 cars are drawn in the picture.)
3. 12 – 6 =, 6 (Make sure 6 eggs are drawn in the picture.)
4. 5 – 3 =, 2 (Make sure 2 people are drawn in the picture.)
5. 12 – 9 =, 3 (Make sure 3 pencils are drawn in the picture.)

### Page 30  Matrix
Make sure all containers in the "full" column are filled. Make sure all containers in the "half empty" column are half empty. Make sure all containers in the "empty" column are empty.

### Page 31  Order
1. a. b
   b. c
   c. Check to make sure all shelves are filled.
2. a. 3, 2, 1
   b. 2, 1, 3
   c. 3, 2, 1

### Page 32  More Problem Solving
1. a. Make sure last jar is full.
   b. Make sure last jar is full.
   c. Make sure last jar is half empty.
2. a. Make sure there are two glasses of water drawn, 1 + 1 = 2
   b. Make sure there is a full glass of water and then an empty glass of water drawn, 1 – 1 = 0

### Page 33  Assessment
1. a. The first picture should be circled.
   b. The first picture should be circled.
   c. The second picture should be circled.
2. a. The first picture should be circled.
   b. The second picture should be circled.
   c. The first picture should be circled.
   d. The second picture should be circled.
3. Four eggs should be drawn.  12 – 8 =, 4
4. 2, 3, 1
5. The first glass should be half full, the second glass half full, and the last glass completely full.

### Page 34  Capacity Experiment
Answers will vary.

| **Name** | **Date** |
|---|---|

**1.** Trace words.

full empty

**2.** Circle the items that are empty.

*#8988 Targeting Math: Measurement*

| **Name** | **Date** |

Color the containers that would hold the most.

| **Name** | **Date** |
| --- | --- |

Create number sentences to solve each problem.

**1.**

The elevator can take 10 people. How many more can fit into the elevator?

| 1 | 0 | – | 3 | = |
| --- | --- | --- | --- | --- |

☐ people

**2.**

Draw the cars.

The parking lot can take 20 cars. If there are 14 cars parked already, how many more cars can park in the parking lot?

| ☐ | ☐ | ☐ | ☐ | ☐ |
| --- | --- | --- | --- | --- |

☐ cars

**3.**

Draw the eggs.

12 eggs can fit into the egg carton. If there are 6 eggs in the carton, how many more can be put into the carton?

| ☐ | ☐ | ☐ | ☐ |
| --- | --- | --- | --- |

☐ eggs

**4.**

Draw the people.

5 people can fit into the car. If there are 3 people in the car, how many more can fit?

| ☐ | ☐ | ☐ |
| --- | --- | --- |

☐ people

**5.**

Draw the pencils.

12 pencils fit into a pencil box. If there are 9 pencils in the box, how many more pencils can fit in?

| ☐ | ☐ | ☐ | ☐ |
| --- | --- | --- | --- |

☐ pencils

(29)

| Name | Date |
|------|------|

**1.** Complete.

**Example:**

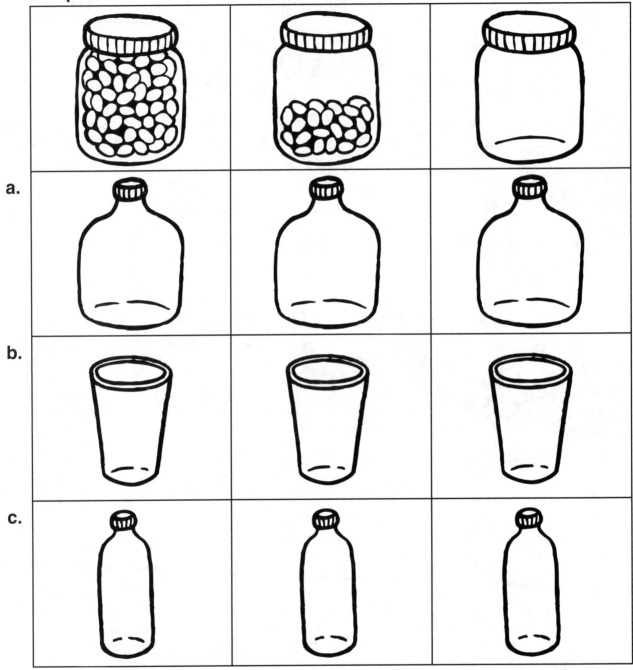

a.

b.

c.

_____    _____    _____

**2.** On the lines below the chart, label the matrix *full*, *half full*, and *empty*.

**Name**                                    **Date**

1. **a.** Which shelf is empty? ☐

   **b.** Which shelf is full? ☐

   **c.** Draw more toys to fill the shelves.

 A

 B

 C

2. Order 1–3 from smallest to largest.

**a.**  ___  ___  ___

**b.**  ___  ___  ___

**c.** 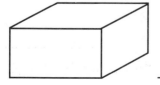 ___  ___  ___

31

| **Name** | **Date** |
|---|---|

1. Predict what you think will happen. Draw your answer. Use water and jars to help you, if necessary.

   **a.** Susan poured half a jar of water into a jar half full of water.

           |

   **b.** Charles poured half a jar of jellybeans into half a jar of jellybeans.

           |

   **c.** Duncan poured half a jar of water into an empty jar.

           |

2. Draw the problem and then write a number sentence. Use water and glasses to help you, if necessary.

   **a.** Peter has a glass full of water, and Mark has a glass full of water. How many glasses of water altogether?

   **b.** Clarissa has a full glass of water and drinks all of the water. How much water is left?

*#8988 Targeting Math: Measurement*                    © *Teacher Created Resources, Inc.*

| **Name** | **Date** |
| --- | --- |

1. Circle the items that are empty.

   **a.**

   **b.**

   **c.**

2. Circle the containers that would hold the most.

   **a.**    **b.**

   **c.**     **d.**

3. Draw to answer this problem.

   12 eggs can fit into the egg carton. If there are 8 eggs in the carton, how many more can be put into the carton? Draw the rest of the eggs.

    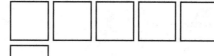  eggs

4. Order from smallest to largest, 1 to 3.

    ____    ____    ____

5. Draw what happens.

   Brandon added half a glass of water to half a glass of water.

    +   =

| Name | Date |
|------|------|

Work in groups of three or four.

## Directions

1. Collect six containers which are different but about the same height.

2. Pour 50 mL of water into each container. Mark the water level with a permanent marker.

3. Keep repeating step 2. Stop when one container is full.

4. Discuss what has happened.

5. Write three sentences about your experiment.

a. _____

b. _____

c. _____

# MORE CAPACITY/ VOLUME

## Unit 2

**Ordering
Estimating
Measuring using
    informal units
Finding differences
Capacity
Graphs
Problem solving**

## Objectives

- *use available technology to help in the solution of mathematical problems*
- *use number skills involving whole numbers to solve problems*
- *order whole numbers up to 999*
- *estimate, compare, and order the capacity of containers using informal units*
- *approximate, count, compare, order, and represent whole numbers and groups of objects up to 100*
- *sort and describe objects in terms of their features such as size and shape*
- *count collections up to 10 and beyond*
- *read and interpret graphs made from objects*
- *describe and model the relationship between the parts and the whole*

## Language

*most, carry, largest, smallest, number, how many more?, least, difference, estimate, fill, fill how many?, actual, order, least, greatest, shared*

## Materials/Resources

*colored pencils, calculators, unit cubes (Base 10), paper cups, water, empty ice-cream containers, plastic containers, small bucket, milk cartons, empty matchboxes, tennis balls, yogurt cups, mugs, buttons, marbles, saucepan, colander, shoebox, bowl*

## Contents of Student Pages

- \* *Materials needed for each reproducible student page*

### Remember

❑ *Encourage students to always estimate first.*

## Additional Activities

❑ Make a collection of containers of different shapes that have similar capacities (and volumes) and allow students to explore volume and capacity.

❑ Encourage students to make a variety of models using the same number of identical blocks. Have them compare their models and discuss the volumes.

❑ Take the students to the local shops. Find as many things as possible that are measured by volume or capacity and record the findings.

❑ Find simple recipes, like the pancake recipe, to use with the students.

❑ Encourage students to bring in 1-liter containers. Make a display.

❑ Link with science and technology and allow students to explore displacement.

❑ Provide opportunities for students to place containers in order of size according to the number of smaller units that fill them.

❑ Have a collection of items (Examples: tissues or eggs). Encourage students to estimate how many in each. Check labeling on items. Students could find which items contain the most/least.

## Answers

### Page 37  Ordering
1. 747
2. Fokker
3. 69
4. Left to Right:  2, 4, 3, 1
5. Concorde

### Page 38  Estimating
1. Teacher to check.
2. Left to right:  3, 2, 4, 1

### Page 39  More Estimating
Teacher to check.

### Page 40  Finding Capacity
Make sure estimates are reasonable.  Below are the actual answers.

1. a. 8
   b. 7
   c. 8
   d. 8
   e. 10
   f. 9
2. Teacher to check.
3. The model in letter "e" should be colored red, and the model in letter "b" should be blue.
4. The following should be linked together: a/f, b/g, c/e, d/h

### Page 41  Interpreting a Graph
1. shoebox
2. smaller saucepan
3. bigger saucepan, colander
4. bigger saucepan, colander, shoe box
5. bowl and smaller saucepan
6. bowl
7. smaller saucepan
8. 5 balls
9. 27
10. 30

### Page 42  Problem Solving
1. 1 cup of flour should be colored.  1/2 teaspoon of baking soda should be colored.  1/2 cup of milk should be colored.  1/4 cup of sugar should be colored.  1 egg should be colored.
2. a. 2 cups of self-rising flour
      1 teaspoon of baking soda
      1 cup of milk
      ½ cup of sugar
      2 eggs
   b. 3 cups of self-rising flour
      1½ teaspoons of baking soda
      1½ cups of milk
      ¾ cups of sugar
      3 eggs
   c. 10
   d. 21

### Page 43  Assessment
1. Teacher to check.
2. Make sure estimates are reasonable.  Below are the actual answers.
   a. 7
   b. 10
   c. 11
   d. 11
   e. 3
   f. 10
3. Teacher to check.

36

| **Name** | **Date** |

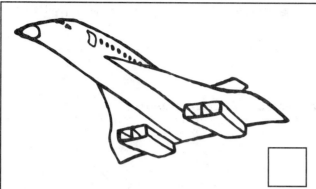

The **Concorde** can carry 128 passengers and travel about 1,450 miles per hour.

The **747** can carry 350 passengers and can travel about 608 miles per hour.

The **Airbus** can carry 281 passengers and can travel about 414 miles per hour.

The **Fokker** can carry 107 passengers and can travel about 497 miles per hour.

1. Which plane can carry the most passengers?_____

2. Which plane carries the least number of passengers? _____

3. Use a calculator to find how many more passengers the 747 can carry than the Airbus. _____

4. In the box provided next to each plane, order the planes 1–4 in order of the number of passengers they can carry, from smallest to largest number.

5. Which plane travels the fastest?_____

37

| **Name** | **Date** |
|---|---|

1. Estimate first. Use a paper cup to fill these containers with water. Find the difference.

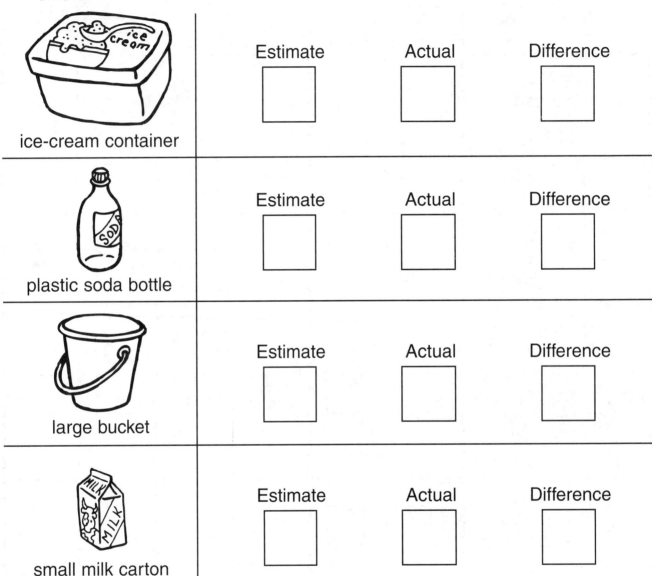

ice-cream container

| Estimate | Actual | Difference |
|---|---|---|
| ☐ | ☐ | ☐ |

plastic soda bottle

| Estimate | Actual | Difference |
|---|---|---|
| ☐ | ☐ | ☐ |

large bucket

| Estimate | Actual | Difference |
|---|---|---|
| ☐ | ☐ | ☐ |

small milk carton

| Estimate | Actual | Difference |
|---|---|---|
| ☐ | ☐ | ☐ |

2. Order the containers 1–4 from least to greatest volume.

 ☐    ☐    ☐    ☐

| Name | Date |
|------|------|

Estimate and then check how many are needed to fill the containers.

| | Base 10 Ones | | Buttons | | Marbles | |
|---|---|---|---|---|---|---|
| | Estimate | Actual | Estimate | Actual | Estimate | Actual |
|  yogurt cup | ☐ | ☐ | ☐ | ☐ | ☐ | ☐ |
|  mug | ☐ | ☐ | ☐ | ☐ | ☐ | ☐ |
|  empty matchbox | ☐ | ☐ | ☐ | ☐ | ☐ | ☐ |
|  ice-cream container | ☐ | ☐ | ☐ | ☐ | ☐ | ☐ |
|  paper cup | ☐ | ☐ | ☐ | ☐ | ☐ | ☐ |

39

| **Name** | **Date** |
|---|---|

1. Estimate and then count the number of cubes that have been used to make each model.

**a.**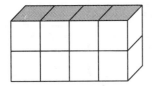

Estimate: ☐

Actual: ☐

**b.**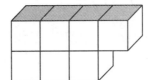

Estimate: ☐

Actual: ☐

**c.**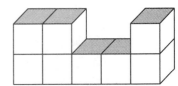

Estimate: ☐

Actual: ☐

**d.**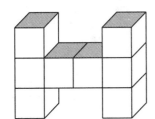

Estimate: ☐

Actual: ☐

**e.**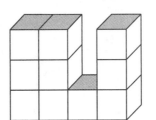

Estimate: ☐

Actual: ☐

**f.**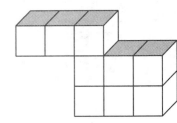

Estimate: ☐

Actual: ☐

2. Make the models above using cubes.

3.  **a.** Color the model that takes up the most space red.

   **b.** Color the model that takes up the least space blue.

4. Look at these groups of cubes.  Link the pairs that have equal numbers.

**a.**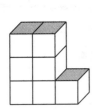
**b.**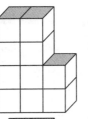
**c.**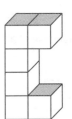
**d.**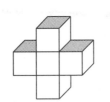

**e.**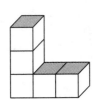
**f.**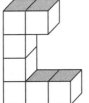
**g.**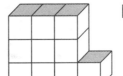
**h.**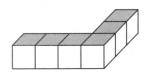

⓵ **40**

**Name**                                    **Date**

Simon used tennis balls to find the volume of different containers. He then drew a graph. Answer the questions below.

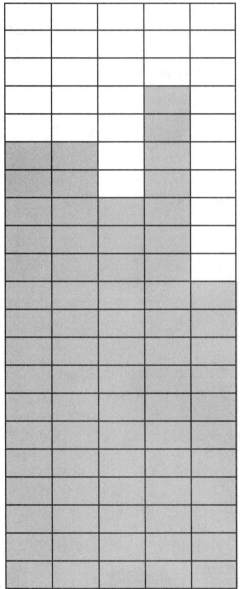

1. Which container has the greatest volume?

   _____

2. Which container has the least volume?

   _____

3. Which two containers have the same volume?

   _____

4. Which containers have a volume of more than 14 tennis balls? _____

5. Which containers have a volume of less than 16 tennis balls? _____

6. Which container has a volume of 14 tennis balls? _____

7. Which container has a volume of 11 tennis balls? _____

8. How many more tennis balls were needed to fill the bigger saucepan than the smaller saucepan? _____

9. How many tennis balls were needed to fill the two saucepans? _____

10. How many tennis balls were needed to fill the colander and the bowl? _____

41

| **Name** | **Date** |
|---|---|

1. Here is a recipe for Scottish pancakes. Color the measurement of each ingredient to show what is needed.

**INGREDIENTS**
- 1 cup of self-rising flour
- ½ teaspoon of baking soda
- ½ cup of milk
- ¼ cup sugar
- 1 egg

**flour**　　　**milk**　　　**sugar**

**DIRECTIONS**
- Sift flour and baking soda.
- Stir in sugar.
- Beat egg lightly and add to mixture.
- Beat in milk.
- Mix to remove all lumps.
- Put spoonfuls into a greased electric frypan.
- When air bubbles appear, turn over with a spatula.
- Remove when cooked.
- Put butter or syrup on pancakes and eat!

**baking soda**　　　**egg**

2. Answer these problems.

   **a.** If I wanted to make a double batch of pancakes, write how much of each ingredient would be needed.

   _____ of self-rising flour

   _____ of baking soda

   _____ of milk

   _____ of sugar

   _____ eggs

   **b.** If I wanted to make three batches of pancakes, write how much of each ingredient would be needed.

   _____ of self-rising flour

   _____ of baking soda

   _____ of milk

   _____ of sugar

   _____ eggs

   **c.** If Julia made 20 pancakes and shared them equally between herself and Amanda, how many pancakes would they each get?

   _____

   **d.** If Brandon ate 8 pancakes, Julian ate 6 pancakes and Aaron ate 7 pancakes, how many pancakes did they eat altogether?

   _____

42

| Name | Date |
|------|------|

1. Find the number of tennis balls needed to fill these containers. Estimate first. Find the difference between the estimate and the actual number.

   | | | Estimate | Actual | Difference |
   |---|---|---|---|---|

   a.  saucepan

   b.  colander

   c. 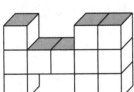 shoebox

2. Estimate and then count the number of cubes.

   a.

   Estimate: ____

   Actual: ____

   b.

   Estimate: ____

   Actual: ____

   c.

   Estimate: ____

   Actual: ____

   d.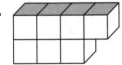

   Estimate: ____

   Actual: ____

   e.

   Estimate: ____

   Actual: ____

   f.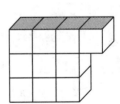

   Estimate: ____

   Actual: ____

3. Estimate and then find how many tennis balls are needed to fill these containers.

    bowl

   Estimate: ____

   Actual: ____

   Estimate: ____

   Actual: ____

# LENGTH

In this section there are three units relating to length. Students demonstrate an understanding of length through activities based on long and short, drawing taller or shorter objects, and comparing lengths. Lengths are ordered from longest to shortest and vice versa and the meanings of words such as wider, thicker, etc., are explored. Estimating measurements is encouraged. Perimeters are measured using centimeters and the yard is introduced. Centimeters are also used to find places on a map.

There are three assessment pages and two activity pages. One activity page involves drawing the longest snake following rules and the other activity asks the children to help Booboo get to the top of the pile of dirt.

# BEGINNING LENGTH

## Unit 1

**Comparing length**
**Ordering length**
**Long and short**
**Wide and narrow**
**Thicker and taller**

## Objectives

- sorts and describes objects in terms of their features, such as size and shape
- explores 2-D shapes and 3-D objects, describing them in comparative language
- finds things bigger or smaller than a given object
- understands everyday comparative language associated with length, mass, and capacity
- responds to and uses everyday comparative language relating to length when recording and communicating measurements
- uses appropriate language for each length, capacity, and mass

## Language

*tall, short, long, wide, thick, narrow, thin, big, little, medium-sized, tallest, shortest, smallest, thickest, longest, order, taller, wider, thicker, longer, height, about the same, as tall as, not as tall as*

## Materials/Resources

*scissors, glue, pencils, colored pencils*

## Contents of Student Pages

* Materials needed for each reproducible student page

### Page 47  Comparing

coloring a particular attribute, such as shorter or wider

* colored pencils

### Page 48  Long and Short

tracing words; drawing lines to match words with objects

* pencils

### Page 49  Ordering Length

longest, shortest; cutting out and matching objects with others; ordering numerical sizes

* scissors, pencils, glue

### Page 50  Drawing Length

drawing particular size attributes

* colored pencils

### Page 51  Drawing More Length

making objects longer and taller

* pencils

### Page 52  Assessment

* colored pencils

### Remember

❏ Make sure students understand the attributes involved with each reproducible, as size attributes are often confused. If they are unsure, provide more activities.

❏ Encourage correct terminology (Example: tall instead of big).

# Additional Activities

- ❑ Provide many opportunities for students to handle, explore, and experiment with a wide variety of materials through stacking, building, and joining activities.

- ❑ Provide a wide variety of sorting materials for children to sort according to size.

- ❑ Encourage students at all times to use their own language to describe their findings.

- ❑ As much as possible, link the concept of length to other key learning areas (Examples: creative arts—provide materials in different lengths for students to use freely in collage; painting—give students long/short paper or wide/narrow paper; drawing—give students thin/thick paper).

- ❑ Cook with the students. Make big/little cakes, long/short snake biscuits, long/short spaghetti.

- ❑ Have a size party. Make some food with the students and get them to bring food from home (Examples: big/little sandwiches, thin/thick sausages).

- ❑ Make a chart about lid sizes with the students. Encourage them to bring in lids. Draw around them. Update it when a student brings in a lid of a different size. Encourage estimation. Is it bigger?

- ❑ Make a shoe shop with the students. Allow time for play. Link with the money strand.

- ❑ There are many picture books that have length as their theme (Examples: Goldilocks and the Three Bears, The Very Hungry Caterpillar by Eric Carle, Titch by Pat Hutchins). Read them to the students. Use them for discussion. Have students illustrate the books. When possible, have them dramatize the story.

- ❑ At music time, get students to walk/run on all fours, on tippy toes, take long steps, baby steps, etc.

- ❑ Encourage the students to dictate or write size stories. Put them in a class book.

- ❑ Get students to find pictures to draw and make class length books (Example: a child draws a big bear and a little bear).

- ❑ Encourage students to find pictures in magazines of objects in different sizes (Examples: letters and numerals). Paste them onto charts.

- ❑ Have students suggest what could make a safe tall tower. Students find what they can use to make the tallest tower.

- ❑ Provide many opportunities for students to compare sizes before introducing informal units for measuring.

# Answers

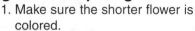

### Page 47  Comparing
1. Make sure the shorter flower is colored.
2. Make sure the wider door is colored.
3. Make sure the thicker tree trunk is colored.
4. Make sure the shorter pencil is colored.
5. Make sure the taller clown is colored.
6. Make sure the smaller ball is colored.

### Page 48  Long and Short
1. Make sure the word *long* is traced. Make sure that the following objects are connected to the word long: longer bead necklace, the balloon with the longer string, the kite with the longer kite string, the longer pencil, the longer scarf, the longer snake, and the longer chain.
2. Make sure the word *short* is traced. Make sure that the following objects are connected to the word short: shorter bead necklace, balloon with the shorter string, the kite with the shorter string, the shorter pencil, the shorter snake, the shorter scarf, and the shorter chain.

### Page 49  Ordering Length
1. Make sure correct shoes are pasted next to each other.
2. The word *longest* should be written under the middle shoe. The word *shortest* should be written under the last shoe.
3. Answer will vary.
4. 3, 5, 7, 9, 10

### Page 50  Drawing Length
1. Make sure a longer flagpole with a flag is drawn.
2. Make sure a tree with a shorter trunk is drawn.
3. Make sure a shorter door is drawn.
4. Make sure a shorter ladder is drawn.
5. Joey, Bozo, Joey

### Page 51  Drawing More Length
Make sure all objects are longer or taller.

### Page 52  Assessment
1. a. big, little
   b. shorter, longer
2. Taller flower should be circled.
3. A taller candle should be drawn.
4. Make sure the train is longer.
5. a. Long stripes should be drawn.
   b. Short stripes should be drawn.

| **Name** | **Date** |
|---|---|

Color the:

1. shorter flower.

2. wider door.

3. thicker tree trunk.

4. shorter pencil.

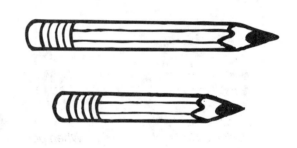

5. taller clown.

6. smaller ball.

                                                     *#8988 Targeting Math: Measurement*

| **Name** | **Date** |
|---|---|

1. Trace over the word *long*, then draw lines to match the longer things to the word.

long

short

2. Trace over the word *short*, then draw lines to match the shorter things to the word.

| **Name** | **Date** |
|---|---|

1. Cut out the shoes at the bottom of the page and paste them next to the correct shoe size.

_____

_____

_____

2. Write *longest* next to the longest pair of shoes.

   Write *shortest* next to the shortest pair of shoes.

3. What is your shoe size? _____

4. Put these shoe sizes in order from *smallest to largest.*

| 9 | 3 | 10 | 7 | 5 |
|---|---|----|---|---|
|   |   |    |   |   |

✂ - - - - - - - - - - - - - - - - - - - - - - - - - - - - - - - - -

49

| **Name** | **Date** |
|----------|----------|

1. Draw a **longer** flagpole with a flag.

2. Draw a tree with a **shorter** trunk.

3. Draw a **shorter** door.

4. Draw a **shorter** ladder.

Joey

Bozo

5.

Who is **taller**? _____

Who has the **longer** hair?

_____

Who has the **longer** shoes?

_____

(50)

| **Name** | **Date** |
|---|---|

1. Make the coat longer.

2. Make the building taller.

3. Make the scarf longer.

4. Make the necklace longer.

5. Make the kite string longer.

| **Name** | **Date** |
|----------|----------|

1. Color the correct word box.

a.

|  | |
|----------------------|----------------------|
| big | big |
| little | little |

b.

| pencil | pencil |
|--------|--------|
| shorter | shorter |
| longer | longer |

2. Circle the flower that is **taller**.

3. Draw a **taller** candle.

4. Make the train **longer**.

5. a. Draw **long** stripes.

b. Draw **short** stripes.

# MORE LENGTH

## Unit 2

Informal units
Ordering length
Comparing lengths
Shortest route
Formal units

## Objectives

- responds to and uses everyday comparative language relating to length when recording and communicating measurements
- describes relative distances using such terms as "nearer" and "further"
- sorts and describes objects in terms of their features such as size and shape
- uses matching to identify longer than, shorter than, and same length objects
- understands everyday comparative language associated with length
- tests estimates of physical quantities using appropriate informal units
- estimates, compares, orders, and measures the length of objects and the distance between them using informal or formal units
- counts collections up to ten objects and beyond
- uses formal units carefully and consistently to measure and compare length
- compares and orders length using indirect methods
- uses guessing and checking for length comparisons of two objects

## Language

shortest, longest, smallest, nearest, furthest, near, taller than, shorter than, tallest, order, as long as, longer, shorter, estimate, actual, longer than, height, same, numbers, length, width, measure

## Materials/Resources

coloring pencils, sharp pencils, inch rulers, informal units (such as counters, buttons, and paper clips), die, large piece of art paper

## Contents of Student Pages

* Materials needed for each reproducible student page

### Page 55 Shortest Way
coloring the shortest route; coloring a longer route; answering questions about size

* colored pencils

### Page 56 Ordering Length
ordering length from pictures; answering questions about length; ordering length by matching

* colored pencils

### Page 57 Comparing Length
drawing objects longer, shorter, taller

* colored pencils

### Page 58 Informal Units
choosing suitable objects to measure the lengths of drawings; opportunities to estimate and check

* counters, buttons, paper clips

### Page 59 Inches
using an inch ruler to measure lengths of objects pictured; ordering according to length

* pencils, inch rulers

### Page 60 More Inches
estimating and checking using an inch ruler; ordering according to length

* pencils, inch rulers

### Page 61 Assessment
* pencils, colored pencils, inch rulers

### Page 62 Activity—Make the Snake
* 1 die, large piece of art paper, ruler with inches, pencil

## Remember

❑ As students are required to count informal and formal units when measuring, they should be given many opportunities to develop counting skills.

❑ Before students are asked to measure using informal or formal units, they need to have many sorting and comparing length experiences.

❑ Focus the students' attention on the need for a common starting point when comparing objects.

(53)

## Additional Activities

❑ Provide many opportunities for students to compare lengths by matching.

❑ Provide many opportunities for students to suggest and use a wide variety of informal units to measure objects in and out of the classroom. Encourage estimate/check method.

❑ Students write their own "length" stories.

❑ Read books to students about length, such as "Jim and the Beanstalk" by Raymond Briggs. Link with creative arts and illustrate aspects of the story.

❑ Link length findings with graph work whenever possible.

❑ Link with creative arts and science and technology. Students make the planets out of papier-mache and suspend from the ceiling.

❑ Look at shadows in the playground and encourage students to find ways of measuring them using informal units.

❑ Decide on a theme and provide materials for students to make objects to be displayed (Examples: flowers of different heights).

❑ As a class, make hand prints on a long piece of paper (make sure they do not touch). Put it up in the classroom to measure the length or width of the room.

❑ Go on a length search. Divide the class into small groups. Each group is given something to find, such as "something as wide as a book" or "something longer than this stick." Students record their findings and then report their findings to the class.

❑ Have competitions. Who can throw the shoes the furthest? Make paper airplanes. Whose plane can go the furthest?

## Answers

### Page 55  Shortest Way

1. a. The shortest way would be a straight line from Sally's house to Kelly house.
   b. Any other way would be longer. Teacher can check.
   c. Answers can vary. Make sure answers list things located on the map.
2. a. Mercury
   b. Mercury
   c. Pluto
   d. Venus, Mars
   e. Make sure moon is drawn near Earth.

### Page 56  Ordering Length

1. a. Jake
   b. Andrew
   c. Andrew, Jake
2. a. Answers left to right:  1, 3, 2
   b. Answers top to bottom:  3, 1, 2
3.

| blue | yellow | red | blue | yellow |
|------|--------|-----|------|--------|

| red | blue | red | yellow |
|-----|------|-----|--------|

### Page 57  Comparing Length

Teacher to check.

### Page 58  Informal Units

Teacher to check.

### Page 59  Inches

1. a. finger = 2 inches, beads = 4 inches, pencil = 3 inches, chain = 2 inches
   b. beads
   c. the finger and the chain
2. a. blocks = 2 inches, building = 4 inches, tree = 3 inches
   b. Answers left to right:  1, 3, 2

### Page 60  More Inches

1. Make sure all estimates are reasonable.  Below are the actual measurements.
   a. 3
   b. 5
2. Make sure all estimates are reasonable.  Below are the actual measurements.
   a. 1
   b. 2
3. a. Make sure all estimates are reasonable.  The following are the actual measurements listed left to right:  2, 1, 3.
   b. 2, 1, 3

### Page 61  Assessment

1. pencil = length, book = width, tree = height
2. 2, 1, 3
3. a. Make sure estimate is reasonable.  The actual answer is 4.
   b. Make sure the truck is 1 inch long.
   c. Make sure the truck is 3 inches long.
   d. Make sure truck "a" is colored.

54

| **Name** | **Date** |
|---|---|

1. Sally lives in Greentown. She wants to visit her friend Kelly.

   **a.** Color the shortest way red.

   **b.** Color a longer way blue.

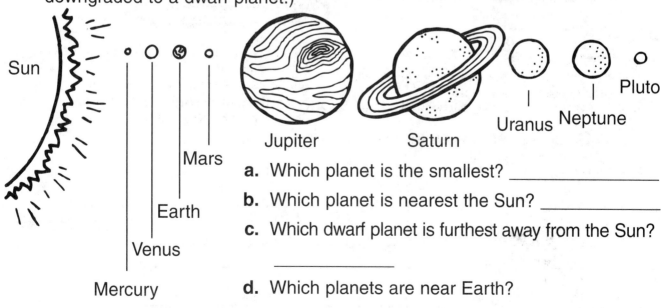

   **c.** What would she pass going the longer way?

   _____

   _____

   _____

2. There are eight planets in the solar system. (In 2006, Pluto was downgraded to a dwarf planet.)

   **a.** Which planet is the smallest? _____

   **b.** Which planet is nearest the Sun? _____

   **c.** Which dwarf planet is furthest away from the Sun?

   _____

   **d.** Which planets are near Earth?

   _____ and _____

   **e.** Draw a moon near Earth.

(55)

| **Name** | **Date** |
|---|---|

1. The children are waiting for the bus.

Andrew    Jake    Paula

    **a.** Who is the tallest? _____

    **b.** Who is the shortest? _____

    **c.** Paula is taller than _____

     but shorter than _____

2. **a.** Order from shortest to tallest.

    Use numbers 1, 2, 3.

   **b.** Order from longest to shortest.

    Use numbers 1, 2, 3.

3. **a.** Color the rectangles that are as long as this one red.

   **b.** Color the rectangles that are longer blue.

   **c.** Color the rectangles that are shorter yellow.

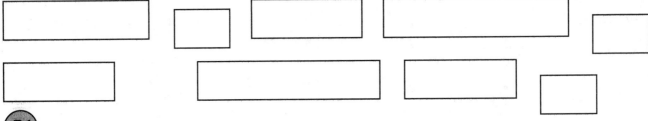

**Name** **Date**

Comparing Length

1. Find and draw something in the room that is:

   **a.** longer than a pencil.

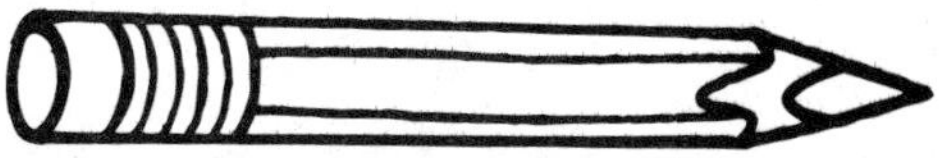

   **b.** shorter than a pencil.

2. Find and draw something in the room that is:

| taller than you. | shorter than you. |
| --- | --- |
| | |

| **Name** | **Date** |
|----------|----------|

Choose suitable objects (Example: counters or buttons) to measure the lengths of these caterpillars. Estimate, then measure and count.

Object _____

Estimate ☐    Actual ☐

Object _____

Estimate ☐    Actual ☐

Object _____

Estimate ☐    Actual ☐

Object _____

Estimate ☐    Actual ☐

Object _____

Estimate ☐    Actual ☐

Object _____

Estimate ☐    Actual ☐

Object _____

Estimate ☐    Actual ☐

Object _____

Estimate ☐    Actual ☐

Object _____

Estimate ☐    Actual ☐

*#8988 Targeting Math: Measurement*

| Name | Date |
|------|------|

1. **a.** Measure the length of each object with an inch ruler. Write the measurement on the line.

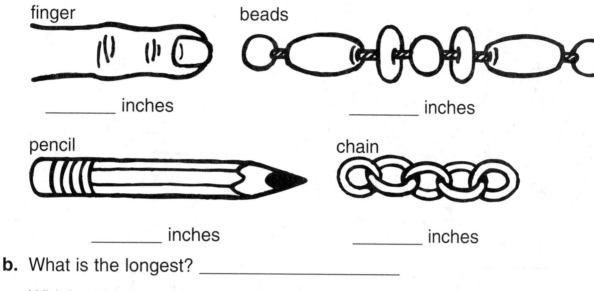

finger

_____ inches

beads

_____ inches

pencil

_____ inches

chain

_____ inches

**b.** What is the longest? _____

**c.** Which objects are the same lengths? _____

2. **a.** Measure the heights of the objects below with an inch ruler. Write the measurement on the line.

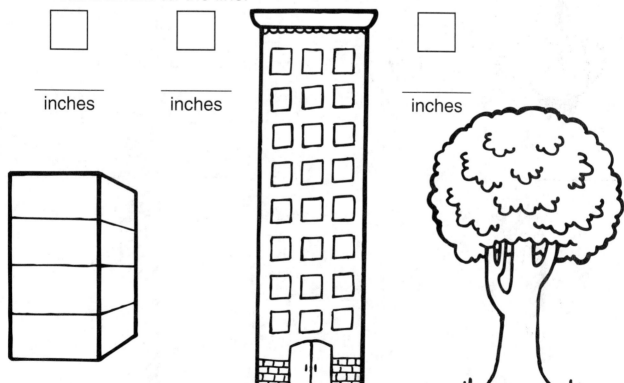

_____
inches

_____
inches

_____
inches

**b.** Order from shortest to tallest. Write the numbers 1, 2, 3 in the boxes near the objects.

59

| Name | Date |

1. Estimate and then measure the **length** of the pencils using an inch ruler.

   **a.**

   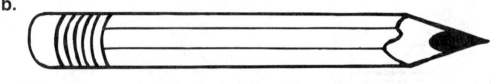

   Estimate: ＿＿＿＿    Actual: ＿＿＿＿

   **b.**

   Estimate: ＿＿＿＿    Actual: ＿＿＿＿

2. Estimate and then measure the **width** of these doors using an inch ruler.

   **a.**

   Estimate: ＿＿＿＿

   Actual: ＿＿＿＿

   **b.**
   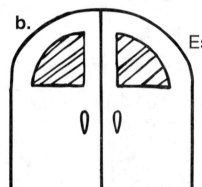
   Estimate: ＿＿＿＿

   Actual: ＿＿＿＿

3. Estimate and then measure the **height** of these candles using an inch ruler.

   **a.**

    Estimate: ＿＿＿＿        Estimate: ＿＿＿＿        Estimate: ＿＿＿＿

   Actual: ＿＿＿＿          Actual: ＿＿＿＿          Actual: ＿＿＿＿

**60** **b.** Order the candles from shortest to tallest in the boxes below the candles. Use numbers 1, 2, 3.

**Name** | **Date**

1. Draw a line to the correct label.

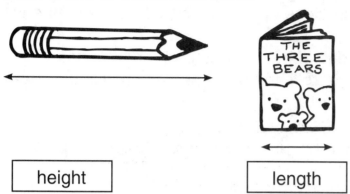
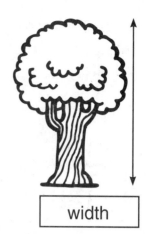

| height | length | width |

2. Order the trucks shortest to longest.  Use numbers 1, 2, 3.

3.  **a.** Estimate then measure the length of the truck using an inch ruler.

Estimate: ☐

Actual: ☐

   **b.** Draw a truck that is 1 inch long.

   **c.** Draw a truck that is 3 inches long.

   **d.** Color the longest truck.

61

**Name**                                          **Date**

**Number of Players:** 2
**Materials:**               1 die, a large piece of art paper each, a ruler, a pencil
**How to Play:**

- On the edge of the paper draw a snake's head.
- Each player rolls the die. The person with the lowest number starts.
- The first person rolls the die.
- Using the ruler, draw that number of inches onto your snake.
- Stop after 10 turns each.
- Measure the snakes. The player with the longest snake is the winner.

# EVEN MORE LENGTH

## Unit 3

*Estimating*
*Measuring*
*Perimeter*
*Meter*
*Ordering length*

## Objectives

- recognizes what features of an object can be measured
- tests estimates of physical quantities using appropriate formal and informal units
- directly and indirectly compares lengths and measures and makes lengths and distances by counting uniform units including centimeters
- chooses the appropriate physical attribute when comparing and measuring things and units which relate to that attribute
- measures perimeter
- uses formal units
- recognizes a yard length
- represents, interprets, and explains mathematical technology, including simple graphs and diagrams

## Language

big, little, tallest, shortest, longer, longest, shorter, measured, centimeters, wing span, estimate, actual, perimeter, shapes, total, order, smallest, largest, tally, distance, informal unit, formal unit, ruler, yard, centimeter, length

## Materials/Resources

colored pencils, sharp pencils, centimeter rulers, long narrow paper that can be cut to one yard lengths, craft sticks, yard sticks, scissors, informal units such as paperclips or counters

## Contents of Student Pages

* Materials needed for each reproducible student page

### Page 65  Centimeters
Estimating and measuring using centimeter rulers

* centimeter rulers

### Page 66  More Centimeters
measuring on a map; following instructions

* centimeter rulers

### Page 67  Perimeters
using centimeters to find perimeter

* centimeter rulers

### Page 68  One Yard
how many objects fit along one yard; finding length words in a word puzzle

* yard sticks, craft sticks, scissors, paper

### Page 69  Ordering Length
ordering the lengths of dinosaurs—shortest to longest

### Page 70  Assessment
* centimeter rulers, colored pencils

### Page 71  Activity—Booboo's Trip

## Remember

❑ Students should have lots of experiences with informal units before centimeters and yards are introduced.

❑ Students should have had some experience in measuring perimeters.

❑ Students need to have had experience with measuring centimeters.

## Additional Activities

❑ *Encourage students to decide what objects and what formal units they would like to use. Share results with the class.*

❑ *Have students create their own informal unit—give it a name. Let students use them for measuring. Discuss the merit of each.*

❑ *Students cut one yard from a long string. They tape the string in the form of a shape onto a large sheet of paper. Who can make the biggest/ smallest shape?*

❑ *Find as many ways possible to measure a yard using informal units. Paste the units down, or record the findings in pictorial form. Encourage estimation first.*

❑ *Provide many opportunities for students to measure using a wide variety of informal and formal units.*

❑ *With the use of reference books, find out the size of the tallest person, dinosaurs, animals, (anything the students are interested in). Have them estimate and then find the actual measurements.*

❑ *Give students experiences with perimeters (e.g., plastic farm animals and fences). Students can make paddocks of different perimeters. Make the paddock bigger/smaller. Students can use geoboards to make shapes with different perimeters.*

## Answers

### Page 65  Centimeters

Make sure all estimates are reasonable. The answers below are the actual answers.

1. a. 7
   b. 3
   c. 7
   d. 6
   e. 12
2. a. e
   b. a and c

### Page 66  More Centimeters

1. a. Desert
   b. Shark Bay
   c. Long John's Cliff
   d. Lookout Point
   e. Two Palms
   f. Shipwreck Bay
2. Pictures will vary.

### Page 67  Perimeters

1. 16
2. 14
3. 16
4. 16

5. 16
6. 18

### Page 68  One Yard

1. a. Teacher to check.
   b. Teacher to check.
2.

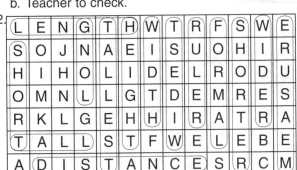

### Page 69  Ordering Length

shortest ↕ longest

| Dinosaur | Length |
| --- | --- |
| Stegosaurus | 10 yards |
| Tyrannosaurus | 13 yards |
| Apotosaurus | 23 yards |
| Brachiosaurus | 31 yards |
| Diplodocus | 33 yards |

### Page 70  Assessment

1. Teacher to check.
2. a. pencil
   b. eraser
3. a. 18 cm
   b. 16 cm
   c. 14 cm
4. The first object should be circled.
5. The second object should be circled.
6. Teacher to check shape and make sure it is 14 centimeters long.

### Page 71  Activity—Booboo's Trip

1. 90 minutes
2. flowers

| **Name** | **Date** |
|---|---|

To find the wing span a butterfly is measured across the widest part.

1. Measure the wing spans of these butterflies using centimeters.  Estimate first.

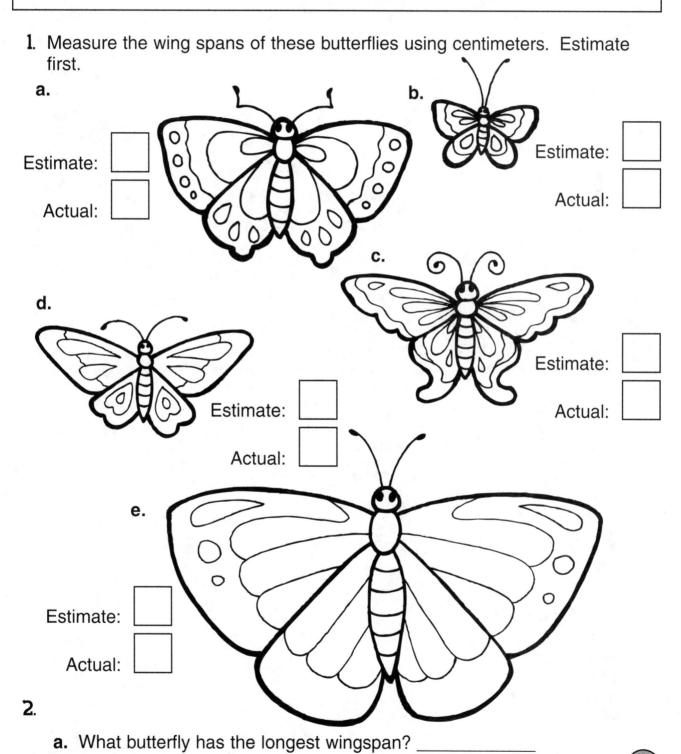

   **a.**

   Estimate: ☐

   Actual: ☐

   **b.**

   Estimate: ☐

   Actual: ☐

   **c.**

   Estimate: ☐

   Actual: ☐

   **d.**

   Estimate: ☐

   Actual: ☐

   **e.**

   Estimate: ☐

   Actual: ☐

2.

   **a.** What butterfly has the longest wingspan? _____

   **b.** Which butterflies have the same wingspan? _____

*#8988 Targeting Math:  Measurement*

| **Name** | **Date** |
|---|---|

You have found a treasure map.  Follow the instructions using a centimeter ruler.

1. Fill in the missing place names.

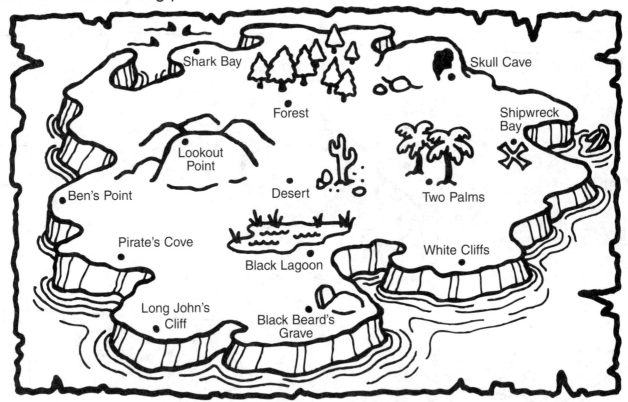

   **a.** From Pirate's Cove, go 5 cm to _____.

   **b.** From Pirate's Cove, go 6 cm to _____.

   **c.** From Pirate's Cove, go 2 cm to _____.

   **d.** From Pirate's Cove, go 3½ cm to _____.

   **e.** From Pirate's Cove, go 9 cm to _____.

   **f.** From Pirate's Cove, go 11 cm to _____. The treasure is here!

2. Draw the treasure.

66

| **Name** | **Date** |
|---|---|

Perimeter is the outer sides of a shape added together.
Example:

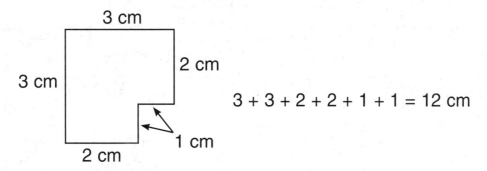

3 + 3 + 2 + 2 + 1 + 1 = 12 cm

Use a centimeter ruler to measure each side of the shape. Add the sides together to find the perimeters of these shapes. Write your answer on the line below each shape.

1.

_____

2.

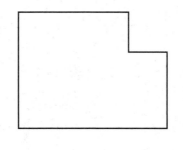

_____

3.

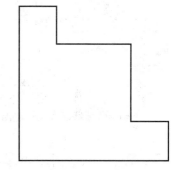

_____

4.

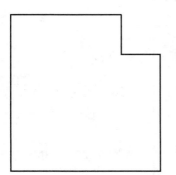

_____

5.

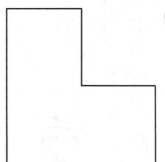

_____

6.

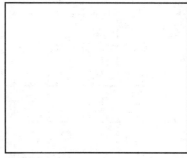

_____

*#8988 Targeting Math: Measurement*

| Name | Date |
|------|------|

1. **a.** Cut a strip of paper to a length of one yard. Estimate, then see how many of each object can fit along one yard.

| Object | | Estimate | Tally | Number |
|--------|--------|----------|-------|--------|
| | How many shoes? | | | |
| | How many hands? | | | |
| | How many craft sticks? | | | |

**b.** You choose two objects to use.

| | | | |
|--|--|--|--|
| | | | |
| | | | |

2. Find and circle the length words in this puzzle.

| L | E | N | G | T | H | W | T | R | F | S | W | E |
|---|---|---|---|---|---|---|---|---|---|---|---|---|
| S | O | J | N | A | E | I | S | U | O | H | I | R |
| H | I | H | O | L | I | D | E | L | R | O | D | U |
| O | M | N | L | L | G | T | D | E | M | R | E | S |
| R | K | L | G | E | H | H | I | R | A | T | R | A |
| T | A | L | L | S | T | F | W | E | L | E | B | E |
| A | D | I | S | T | A | N | C | E | S | R | C | M |
| I | N | F | O | R | M | A | L | U | N | I | T | S |

RULER          FORMAL

SHORTER    LENGTH

WIDTH          SHORT

HEIGHT        TALL

LONG            DISTANCES

WIDER          MEASURE

WIDEST        INFORMAL
                        UNITS

TALLEST

| Name | Date |
|------|------|

## Dinosaurs

Dinosaurs once roamed the earth.

Diplodocus was about 33 yards in length.

Stegosaurus was about 10 yards in length.

Brachiosaurus was about 31 yards in length.

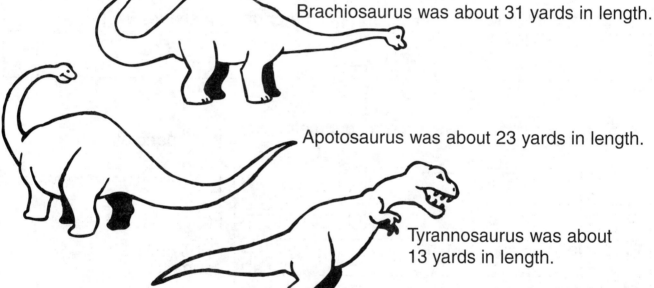

Apotosaurus was about 23 yards in length.

Tyrannosaurus was about 13 yards in length.

Complete the grid.

shortest

↕

longest

| Dinosaur | Length |
|----------|--------|
|          |        |
|          |        |
|          |        |
|          |        |
|          |        |

69

| **Name** | **Date** |
|---|---|

1. Color the two envelopes that are about the same length.

2. The pencil sharpener was 6 centimeters long. The pencil was 17 centimeters long. The eraser was 5 centimeters long.

   a. What is the longest?

   _____

   b. What is the shortest?

   _____

3. Use a centimeter ruler to find the perimeter of each shape. Write your answer inside the shape.

   **a.**

   perimeter _____

   **b.**

   perimeter
   _____

   **c.**

   perimeter
   _____

4. Circle the object with the greater perimeter.

5. Circle the object with the smaller perimeter.

 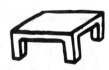

6. Measure to make this shape 14 centimeters long.

| **Name** | **Date** |
|---|---|

Poor old Booboo, a very plump worm, tried to climb a pile of dirt. He wanted to see what was on the other side.

In the first 10 minutes, he climbed 3 cm. Then he needed to rest for 5 minutes. While he was resting, he slipped back 1 cm. He really did want to get to the top. He kept going in the same way—climbing 3 cm in 10 minutes then slipping back 1 cm while he rested for 5 minutes.

1. How long did it take him to reach the top?_____

2. What do you think he saw on the other side? _____

_____

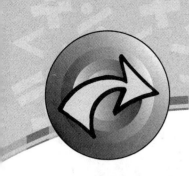

# MASS

These units contain exercises to reinforce the concept that size does not always relate directly to mass. Skills include measuring, estimating, approximating, and comparing and ordering masses. Students identify lighter and heavier, make scales balance, and write mass stories from pictures. Problem solving uses mass in real-life situations and also involves graphs. Two assessment pages and an activity page, on which students house the pet chickens, are included.

# BEGINNING MASS

## Unit 1

**Comparing masses
Lighter and heavier
Scale balance
Counting
Problem solving**

## Objectives

- sorts and describes objects in terms of their features such as size and shape
- estimates, compares, orders, and measures the mass of objects using informal units
- engages in writing texts with the intention of conveying an idea or message
- uses number skills involving whole numbers to solve problems
- answers mathematical questions using objects, pictures, imagery, actions, or trial and error

## Language

greater, mass, easy, hard, hardest, pull, lift, pull up, heavier, lighter, seesaw, lighter than, heavier than, balance, the same as, number, add, take away

## Materials/Resources

sharp pencils, colored pencils

## Contents of Student Pages

### Remember

❏ Give students many opportunities to lift and to use the balance scale so they realize that size is not directly linked to mass.

❏ Encourage students to estimate and then check.

## Additional Activities

❑ *Provide students with many opportunities to lift objects of different sizes and masses. Provide opportunities for lots of discussion and encourage students to record results.*

❑ *Provide opportunities for students to compare and order masses. Record some of the results pictorially.*

❑ *Provide opportunities for students to use the balance scale using many different materials.*

❑ *Take the students to the local park and experiment with a seesaw.*

❑ *Ensure students are involved in a variety of everyday activities where they can experience differences in mass (e.g., moving chairs, piles of books). Encourage discussion of their observations.*

❑ *In P.E. provide opportunities to throw objects with small masses (e.g., feathers). Record the distances. Then throw objects with greater mass. Compare the differences in the effort required.*

❑ *If you have access to a wheelbarrow, allow students to push an empty wheelbarrow and then put objects inside of it and notice the difference. If you have toy trucks for the block area, let students experiment with a full and empty truck. If you have metal trucks, allow students to use in the sand pit for experimentation.*

❑ *Fill identical containers (e.g., ice cream containers or margarine tubs) with different materials (e.g., sand, feathers, corks, pebbles). Allow students to lift them and to put them into order of mass.*

❑ *Students can dramatize pushing and pulling heavy and light objects. Read to the students "The Tale of the Turnip" by Brian Alderson and get them to dramatize the story.*

❑ *Have a collection of items. Choose one and have students choose another item that is lighter than (or heavier than). Check using the balance scale.*

❑ *Get the students to make identical-sized parcels, with identical wrapping, but with different masses. By lifting, students can put them into order.*

❑ *Read to students and discuss Henry Pluckrose's "Weight" (Franklin Watts, London, 1987).*

## Answers

### Page 75  Greater Mass

1. The bigger bear should be circled.
2. The bigger butterfly should be circled.
3. The bigger book should be circled.
4. The bigger truck should be circled.
5. The tire should be circled.
6. The purse with the money in it should be circled.

7. The sweater should be circled.
8. The boot should be circled.

### Page 76  Comparing Mass

1. a. The empty cart should be circled.
   b. The empty toy box should be circled.
   c. The empty wheelbarrow should be circled.
2. a. The pot with the flowers should be circled.
   b. The basket with clothes in it should be circled.
3. The largest fish should be circled.

### Page 77  Heavier/Lighter

1. a. The man should be circled.
2. Make sure the words are traced over.
3. There are three equally balanced seesaws that should be circled.
4. Make sure objects on seesaws are appropriate.

### Page 78  Heavier/Lighter/Same As

1. a. heavier than     c. lighter than
   b. the same as     d. heavier than
2. Make sure sentence matches the picture.

### Page 79  Equal Mass

1. Five marbles should be drawn in the empty bucket.
2. Nine marbles should be drawn in the empty bucket.
3. Six marbles should be drawn in the empty bucket.
4. Eleven marbles should be drawn in the empty bucket.
5. No marbles should be drawn in the empty bucket.
6. One marble should be drawn in the empty bucket.
7. Three marbles should be drawn in the empty bucket.
8. Seven marbles should be drawn in the empty bucket.
9. Ten marbles should be drawn in the empty bucket.
10. Thirteen marbles should be drawn in the empty bucket.

### Page 80  Problem Solving

1. The red bucket should be heavier.
2. The blue bucket should be heavier.
3. The red bucket should be heavier.
4. The blue bucket should be heavier.
5. The red bucket should be heavier.

### Page 81  Assessment

1. a. The stroller with three children in it should be circled.
   b. The picture of three bricks should be circled.
2. a. The child at the bottom of the seesaw should be circled.
   b. The box at the top of the seesaw should be circled.
3. a. the same as
   b. lighter than
4. a. Five marbles should be drawn in the empty bucket.
   b. Seven marbles should be drawn in the empty bucket.
5. The red bucket should be heavier.

**Name**                                    **Date**

Color the one with the greater mass.

*#8988 Targeting Math:  Measurement*

| **Name** | **Date** |
|---|---|

**1.** Circle the objects that are easier to push.

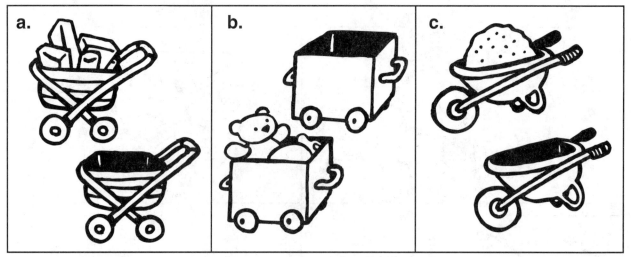

**2.** Circle the objects that are harder to lift.

**3.** Circle the fish that is hardest to pull up.

76

| **Name** | **Date** |
|---|---|

**1.** Circle the heavier person.

**2.** Trace over the words.

lighter          heavier          balance

**3.** Circle the seesaws where the objects are balanced.

  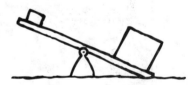

 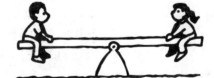 

**4.** Draw objects on each seesaw.

*#8988 Targeting Math: Measurement*

| **Name** | **Date** |
|----------|----------|

**1.** Circle the correct words.

**a.** Peter is (heavier than / lighter than / the same as) Anna.

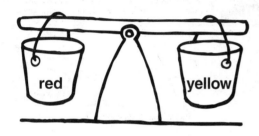

**b.** The red bucket is (heavier than / lighter than / the same as) the yellow bucket.

**c.** The marbles are (heavier than / lighter than / the same as) the book.

**d.** The two boxes are (heavier than / lighter than / the same as) the one box.

**2.** Write a sentence about the picture.

_____

_____

_____

_____

| **Name** | **Date** |
|---|---|

Draw the marbles to balance each bucket.

1.

2.

3.

4.

5.

6.

7.

8.

9.

10.

79

**Name**                                                    **Date**

Draw what would happen.

1. Add 3 marbles to the red bucket.

2. Add 10 marbles to the blue bucket.

3. Take away 4 marbles from the blue bucket.

4. Take away 4 marbles from the red bucket.

5. Add 8 marbles to the red bucket.

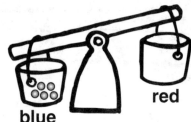

| **Name** | **Date** |
|---|---|

1. Circle the object in each group that would have the greater mass.

**a.**  **b.**

2. **a.** Circle the heavier child. **b.** Circle the lighter box.

3. Circle the correct words.

**a.**

The blue bucket is (heavier than / lighter than / the same as) the green bucket.

**b.**

The doll is (heavier than / lighter than / the same as) the bear.

4. Draw the number of marbles needed to balance each bucket.

**a.**  **b.**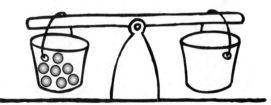

5. Draw what would happen if I put 3 marbles into the red bucket.

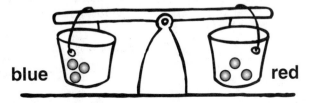

# MORE MASS

## Unit 2

**Estimating
Measuring mass
Ordering mass
Graphs
Pounds
Problem solving**

## Objectives

- estimates, compares, orders, and measures the mass of objects using informal units
- demonstrates an understanding that numbers can be represented using groups of 10, 100, and 1,000
- orders whole numbers to 999
- comments on information in displays produced by himself/herself or others
- makes block graphs using one-to-one correspondence between real data and a representation
- counts collections up to 10 objects and beyond
- makes judgments of comparison of mass by lifting
- uses number skills involving whole numbers to solve problems
- describes and models the relationships between the parts and the whole
- models numbers and number relationships in a variety of ways, and uses them in solving number problems

## Language

heavier, lighter, lightest, heaviest, kilogram, smallest mass, order, greatest mass, mass, balance, equal arm balance, similar mass, scale, push, heft, light, less, same, pull, same mass, graph, estimate, actual, altogether, weigh, total mass

## Materials/Resources

colored pencils, unit cubes (Base 10), calculator, objects to weigh, items to balance (e.g., shells, craft sticks, marbles, thread reels, buttons, bolts, lids), equal arm balances

## Contents of Student Pages

* Materials needed for each reproducible student page

### Page 84  Modeling Mass
using unit cubes; ordering from smallest mass to greatest mass

* unit cubes, colored pencils

### Page 85  Mass Questions
answering questions from given information; completing a find-a-word

### Page 86  Mass Graphing
coloring in squares using given information; answering questions; creating a question

* colored pencils

### Page 87  Estimating
choosing objects to weigh; choosing items to use as balances

* equal arm balance, objects (e.g., plastic cubes, shells, craft sticks, marbles)

### Page 88  Problem Solving
reading mass problems and solving; writing a problem

* calculator

### Page 89  Assessment
* colored pencils

### Page 90  Activity—The New Chicken House
solving a problem

## Remember

- ❏ Link with other math strands.
- ❏ Be sensitive to students' feelings regarding their weight.
- ❏ Provide many types of experiences where comparison of mass is made so students realize that quantity is not linked to mass.

82

## Additional Activities

- ❑ Give students opportunities to estimate and then check the mass of different objects using informal units of measurement (e.g., marbles, plastic cubes).

- ❑ Provide opportunities for students to become familiar with materials that weigh pounds and come in packages weighing pounds (e.g., sugar, rice, flour).

- ❑ If possible, go on an excursion to the local shops. Students guess which shops weigh things using mass. Visit the post office, butcher, grocery store, or a bakery. Each group can report on their findings.

- ❑ If students are interested, find more mass records in the _Guinness Book of World Records._ Discuss and record.

- ❑ If you do not have a student with a weight problem, weigh everyone in the class and record their masses. Use the results to create a graph.

- ❑ Provide opportunities for students to set up different types of shops where scales are used (e.g., post office).

- ❑ Allow students to choose two different items at a time (e.g., buttons and marbles). They decide on a number (e.g., 15) and estimate which item will be heavier.

## Answers

### Page 84  Modeling Mass

1. a. 6 hundred blocks, 3 ten blocks, and 5 one blocks should be colored.
   b. 7 ten blocks and 7 one blocks should be colored.
   c. 1 hundred block, 9 ten blocks, and no one blocks should be colored.
   d. 1 hundred block, 6 ten blocks, and no one blocks should be colored.
   e. 3 hundred blocks, 1 ten block, and no one blocks should be colored.
   f. 2 hundred blocks, 6 ten blocks, and 5 one blocks should be colored.
   g. 2 ten blocks and 1 one block should be colored.
2. 7, 2, 4, 3, 6, 5, 1
3. a. John Brower Minnoch, the heaviest man
   b. well-fed cat

### Page 85  Mass Questions

1. a. book
   b. pencil
   c. 32
   d. 48
   e. 5
   f. 19
   g. 16
   h. scissors and a pencil

2.

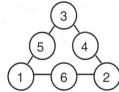

### Page 86  Mass Graphing

1. a. orange
   b. strawberry
   c. lemon and mandarin
   d. passion fruit, strawberry
   e. strawberry
   f. banana
2. a. 64 marbles   b. 46 marbles   c. 9 marbles
3.

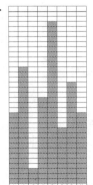

4. Questions will vary.

### Page 87  Estimating

Teacher to check.

### Page 88  Problem Solving

1. a. 4 pounds      e. 97 pounds      i. 10
   b. 26 pounds     f. 9 pounds       j. 5 pounds
   c. $6            g. 78 pounds      k. 20 pounds
   d. 50 pounds     h. 65 pounds
2. They are the same.
3. Answers will vary.

### Page 89  Assessment

1. a. 5 ten blocks and 9 one blocks need to be colored.
   b. 3 hundred blocks, 8 ten blocks, and 5 one blocks need to be colored.
2. a. 6       c. pen      e. 8
   b. ruler   d. 14
3. a. 62 pounds    b. 10 pounds    c. $6

### Page 90  Activity—The New Chicken House

```
        (3)
      (5) (4)
    (1) (6) (2)
```

| **Name** | **Date** |
|---|---|

1. Color to make the numbers.

   **a.** John Brower Minnoch, the heaviest man – 635 pounds

   **b.** St. Bernard dog – 77 pounds

   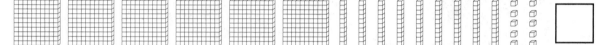

   **c.** Indian tiger – 190 pounds

   **d.** Ostrich – 160 pounds

   **e.** Gorilla – 310 pounds

   **f.** Siberian tiger – 265 pounds

   **g.** a very well-fed cat – 21 pounds

2. In the boxes above, order from smallest to greatest mass using numerals 1–7.

3. **a.** What had the greatest mass? _____

   **b.** What had the smallest mass? _____

| **Name** | **Date** |
|---|---|

1. Alex used the equal arm balance to find the mass of a book, a pencil, and a pair of scissors. This is what he found.

**scissors**

| 58 | craft sticks |
| 23 | lids |
| 24 | thread reels |
| 32 | buttons |

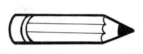

**pencil**

| 14 | craft sticks |
| 4 | lids |
| 5 | thread reels |
| 8 | buttons |

**book**

| 147 | craft sticks |
| 57 | lids |
| 60 | thread reels |
| 80 | buttons |

Answer the questions.

a. Which object was the heaviest? _____

b. Which object was the lightest? _____

c. How many buttons balanced the scissors? _____

d. How many more buttons were needed to balance the book? _____

e. How many thread reels were needed to balance the pencil? _____

f. How many more thread reels were needed to balance the scissors? ___

g. How many lids would be needed to balance 4 pencils? _____

h. Which two items would have a mass of 27 lids? _____ and _____

2. Color these mass words in the puzzle.

HEAVY        MASS
POUND        LESS
SCALE        BALANCE
HEAVIER      HEAVIEST
GREATER      PULL
PUSH         LIGHTER
LIFT         EQUAL
EQUAL ARM    SAME
LIGHT        SMALLER

| K | I | L | P | O | U | N | D | M | A | S | A | M | E | P | A |
|---|---|---|---|---|---|---|---|---|---|---|---|---|---|---|---|
| G | L | I | G | H | T | B | A | S | E | C | M | A | H | U | U |
| R | A | G | P | U | L | E | S | S | M | M | O | P | E | L | R |
| E | H | H | E | A | V | Y | S | C | H | E | Q | U | A | L | E |
| A | S | T | F | I | L | C | D | A | B | R | M | H | U | H | L |
| T | U | E | C | E | C | N | A | L | A | B | A | V | I | L | L |
| E | P | R | R | E | I | V | A | E | H | O | S | I | E | A | A |
| R | H | E | A | V | I | E | S | T | P | A | S | E | S | S | M |
| E | Q | U | A | L | A | R | M | M | U | Q | E | T | T | S | S |

85

| **Name** | **Date** |
| --- | --- |

1. Amanda used marbles to find the mass of some pieces of fruit.

   lemon (L) = 14 marbles
   apple (A) = 23 marbles
   strawberry (S) = 3 marbles
   kiwi fruit (K) = 17 marbles
   orange (O) = 32 marbles
   passion fruit (P) = 11 marbles
   banana (B) = 20 marbles
   mandarin (M) = 14 marbles

   Answer these questions.

   **a.** Which fruit has the greatest mass?

   _____

   **b.** Which fruit has the least mass?

   _____

   **c.** Which two pieces of fruit have the same mass? _____ and

   _____

   **d.** Which fruits have a mass of less than 14 marbles? _____ and _____

   **e.** Which piece of fruit has a mass of 3 marbles?_____

   **f.** Which piece of fruit has a mass of 20 marbles? _____

| L | A | S | K | O | P | B | M |
| --- | --- | --- | --- | --- | --- | --- | --- |

2. **a.** What would be the mass of 2 oranges? _____

   **b.** What would be the mass of 2 apples?_____

   **c.** What would be the mass of 3 strawberries? _____

3. Make a graph by coloring the spaces.

4. Write a question for your graph.

   _____

   _____

   _____

| **Name** | **Date** |
|---|---|

1. Choose any four objects in the room that will fit in the equal arm balance.

    **a.** I choose _____ , _____ , _____ and _____ .

    **b.** By lifting, estimate which is the heaviest.

    The _____ is the heaviest.

    **c.** By lifting, estimate which is the lightest.

    The _____ is the lightest.

    **d.** Check by using the equal arm balance.

    What was the heaviest? _____

    What was the lightest? _____

2. Find the mass of two objects using informal units. Choose three items such as plastic cubes, shells, craft sticks, or marbles. Estimate first.

| Object _____ | | | Object _____ | | |
|---|---|---|---|---|---|
| | Estimate | Actual | | Estimate | Actual |
| Item _____ | ☐ | ☐ | Item _____ | ☐ | ☐ |
| | Estimate | Actual | | Estimate | Actual |
| Item _____ | ☐ | ☐ | Item _____ | ☐ | ☐ |
| | Estimate | Actual | | Estimate | Actual |
| Item _____ | ☐ | ☐ | Item _____ | ☐ | ☐ |

3. **a.** Which was the heaviest object you weighed? _____

    **b.** Which was the lightest object you weighed? _____

    **c.** What did you discover about the items you used to balance the objects?

    _____

    _____

    _____

87

| **Name** | **Date** |
|---|---|

1. Use a calculator to solve these problems.

   **a.** Belinda has a mass of 23 pounds and Karen has a mass of 27 pounds.
   How much more mass has Karen than Belinda?　　_____

   **b.** Amanda's sister weighs 61 pounds and her brother weighs 87 pounds.
   How much heavier is Amanda's brother?　　_____

   **c.** If 1 pound of sugar costs $2, how much would 3 pounds cost? _____

   **d.** If the combined mass of Adam and Ben is 97 pounds and Adam weighs
   47 pounds, how many pounds does Ben weigh?　　_____

   **e.** Katrina, Christie, and Yvette have masses of 29 pounds, 32 pounds and
   36 pounds.  What is their total mass?　　_____

   **f.** Reece went to the supermarket and bought 4 pounds of sugar, 1 pound
   of rice, 2 pounds of plain flour, and 2 pounds of laundry soap.  How
   many pounds altogether?　　_____

   **g.** David weighs 93 pounds and wants to lose 15 pounds.  What weight
   does he want to be?　　_____

   **h.** Clare decided to eat healthy.  If her new weight is 56 pounds and she
   lost 9 pounds, what was her original weight?　　_____

   **i.** If five big potatoes weigh 1 pound and Lisa needs two pounds, how
   many big potatoes does she need altogether?　　_____

   **j.** Deana went to the library and borrowed some books.  Two books
   weighed 1 pound each, 1 book weighed 2 pounds and three books
   weighed 1 pound altogether.  What was the total mass of all the books
   she had to carry to the car?　　_____

   **k.** The baker's sack of flour weighed 40 pounds.  If he used half of the
   flour, how much flour was left?　　_____

2. Trick question—What is heavier:  30 pounds of feathers or 30 pounds of
   potatoes?　　_____

3. Write your own mass problem.

   _____

   _____

**88**

| **Name** | **Date** |
|---|---|

1. Color these groups to make the numbers.

   **a.** A large lizard weighs 59 pounds.

   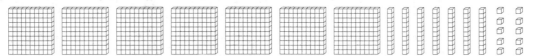

   **b.** A large tortoise weighs 385 pounds.

   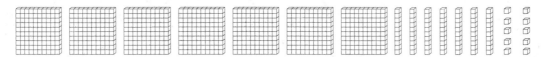

2. Robert measured the mass of a ruler, an eraser, and a pen.  He found:

| **ruler** | **eraser** | **pen** |
|---|---|---|
| 13 lids<br>or 14 thread reels<br>or 19 craft sticks | 6 lids<br>or 7 thread reels<br>or 9 craft sticks | 3 lids<br>or 4 thread reels<br>or 4 craft sticks |

   **a.** Robert found that the eraser was equal to how many lids?_____

   **b.** What had the greatest mass?                          _____

   **c.** What had the least mass?                              _____

   **d.** Robert found that the ruler was equal to how many thread reels? _____

   **e.** How many craft sticks would balance two pens on the equal arm
   balance scale?                                             _____

3. Solve these problems.

   **a.** Joanne, Michelle, and Sara have masses of 20 pounds, 23 pounds, and
   19 pounds.  What is their total mass?                      _____

   **b.** If Mark has a 20-pound bag of cement and uses half of it, how many
   pounds does he have left?                                  _____

   **c.** If 2 pounds of rice costs $3, how much do 4 pounds cost? _____

(89)

| **Name** | **Date** |
| --- | --- |

Sue and Ben keep pet chickens.  These are their chickens.  Each one is a different weight.

Dad built a chicken house so that each one has its own place.  To make the house safe, the chickens have to be arranged so that the chickens along each side weigh a total of nine pounds.  Show how they are arranged.

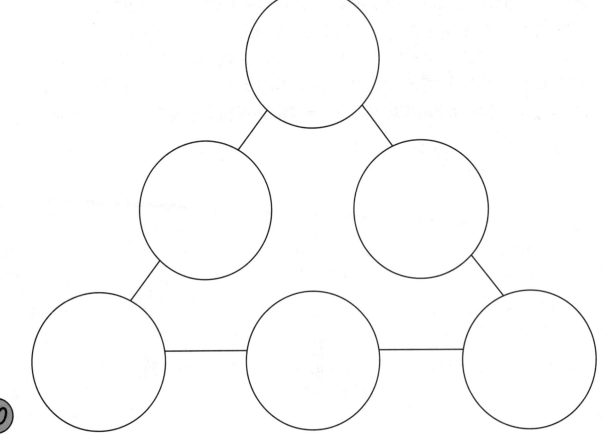

# TIME

This section on time contains exercises in identifying early and late, long time and short time, and night and day. Naming and ordering the days of the week are practiced. Seasons are identified and months of the year sequenced. A calendar is used to solve problems. Time is related to everyday experiences and written words. Analog and digital clocks are used to tell the time in hours, half-hours, and five-minute intervals. Students draw on knowledge to calculate small time intervals and relate time to everyday situations. There are two assessment pages included. Logic is used to solve a time problem on the activity page.

# BEGINNING TIME

## Unit 1

*Days*
*Months*
*Seasons*
*Hours/minutes/seconds*

## Objectives

- responds to or uses everyday language associated with time
- uses everyday language to compare the duration of events
- relates days of the week or months of the year to events in his or her own life
- spells correctly and orders days of the week
- orders days of the week or months of the year
- recognizes that daily activities are related to and influenced by time as indicated by seasons
- counts informal units to measure a period of time
- understands that clocks are used to reflect the passage of time

## Language

after, April, August, autumn, before, coldest, day, December, early, February, fifth, first, Friday, fourth, hottest, hour, January, July, June, last, late, long, longest, March, May, minute, Monday, months, morning, night, November, October, order, Saturday, seasons, second, September, short, shortest, sixth, spring, summer, Sunday, sunrise, sunset, third, Thursday, time, Tuesday, Wednesday, weekend, winter

## Materials/Resources

writing/drawing materials, colored pencils

## Contents of Student Pages

* Materials needed for each reproducible student page

### Page 94 Comparisons
comparing long and short
* colored pencils

### Page 95 Night and Day
identifying night-time and daytime activities; ordering activities
* colored pencils

### Page 96 Days of the Week
ordering days of the week; spelling the days of the week

### Page 97 Months of the Year
ordering months; identifying activities in different months
* colored pencils

### Page 98 Seasons
naming seasons; matching months to seasons

### Page 99 Time Taken
understanding the concept of hours, minutes, and seconds
* colored pencils

### Page 100 Assessment

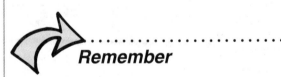

## Remember

❑ Relate classroom activities to the passing of time whenever possible.
❑ Use a variety of words when discussing time.

**92**

## Additional Activities

❑ *Investigate which cultures view Monday as being the first day of the week.*

❑ *Write about your favorite day, month, or season of the year.*

❑ *Make a graph of birthdays of students in the class.*

❑ *Complete a time line of activities you do on a school day or weekend.*

❑ *Write or draw about when you went somewhere and you were either early or late. What happened?*

## Answers

### Page 94 Comparisons

1. a. Circle: plane flying, sleeping, watching TV
   Color: brushing teeth, jumping, eating
2. a. shorter
   b. shorter
   c. longer
3. Teacher to check.

### Page 95 Night and Day

1. a. night
   b. day
   c. day
   d. night
2. a. yes
   b. yes
   c. yes
   d. no
3. Teacher to check.

### Page 96 Days of the Week

1. a. Sunday
   b. Monday
   c. Tuesday
   d. Wednesday
   e. Thursday
   f. Friday
   g. Saturday
2. a. Sunday
   b. Tuesday
   c. Monday
   d. Saturday, Sunday
3. a. Wednesday
   b. Monday
   c. Sunday
   d. Saturday
   e. Tuesday
   f. Thursday
4. Sunday 1, Wednesday 4, Saturday 7, Friday 6, Monday 2, Tuesday 3, Thursday 5

### Page 97 Months of the Year

1. January 1, June 6, February 2, September 9, March 3, August 8, December 12, April 4, May 5, November 11, July 7, October 10

2. a. No
   b. Yes
   c. No
   d. No
   e. Yes
   f. Yes
3. a. Teacher to check.
   b. Teacher to check.

### Page 98 Seasons

1. summer, autumn, winter, spring
2. Autumn = September, October, November
   Summer = June, July, August
   Winter = December, January, February
   Spring = March, April, May
3. a. summer
   b. winter
   c. winter
   d. Teacher to check.
   e. Teacher to check.
   f. Teacher to check.
   g. spring
   h. spring

### Page 99 Time Taken

1. Teacher to check.
2. Teacher to check.
3. Color: clapping, saying "thank you," jumping

### Page 100 Assessment

1. a. shorter
   b. longer
2. a. Sunday
   b. Monday
   c. Tuesday
   d. Wednesday
   f. Thursday
   g. Friday
   h. Saturday
3. a. January
   b. February
   c. March
   d. April
   e. May
   f. June
   g. July
   h. August
   i. September
   j. October
   k. November
   l. December
4. a. June, July, and August should be circled red.
   b. December, January, and February should be circled blue.
   c. March, April, and May should be circled green.
   d. September, October, and November should be circled brown.
5. a. Teacher to check.
   b. Teacher to check.

*#8988 Targeting Math: Measurement*

| **Name** | **Date** |
|---|---|

1. Circle three things that would take the longest time to do. Then color three things that would take the shortest time to do.

2. Complete these sentences using *longer* or *shorter*.

   **a.** It takes a _____ time to brush your hair than to read a book.

   **b.** It takes a _____ time to eat dinner than to walk to school.

   **c.** It takes a _____ time to write a story than to write your name.

3. Draw 3 things that take a long time to do.

| **Name** | **Date** |
|---|---|

1. Write *night* or *day*.

    **a.** We sleep at            _____ .

    **b.** We go to school during the    _____ .

    **c.** We eat lunch during the      _____ .

    **d.** The moon comes out at      _____ .

2. Answer *yes* or *no*.

    **a.** Sunrise is in the morning.    _____

    **b.** Sunset is at the end of the day. _____

    **c.** The opposite of night is day.  _____

    **d.** The sun shines at night.     _____

3. Draw 3 things you can do:

| at night-time | in daytime |
|---|---|
| | |

| **Name** | **Date** |
|---|---|

**1.** Write the days of the week in order starting with Sunday.

**a.** _____      **d.** _____      **g.** _____

**b.** _____      **e.** _____

**c.** _____      **f.** _____

**2.** Answer these.

    **a.** What is the first day of the week?         _____

    **b.** What is the third day of the week?       _____

    **c.** What is the first day of the school week?   _____

    **d.** Which days are the weekend? _____   _____

**3.** Fill in the missing letters.

    **a.** W _ d _ _ sd _ y          **b.** M _ _ d _ _ _

    **c.** _ _ nda _               **d.** _ at _ _ d _ _ _

    **e.** T _ _ _ _ ay             **f.** _ _ urs _ _ _

**4.** Order the days of the week.

Sunday ___1___

Wednesday _____      Friday _____      Tuesday _____

Saturday _____      Monday _____      Thursday _____

    

| **Name** | **Date** |
|---|---|

1. Order the months of the year.

January ___1___       March _____       May _____

June _____       August _____       November _____

February _____       December _____       July _____

September _____       April _____       October _____

2. Answer *Yes* or *No*.

   a. March is the fifth month of the year. _____

   b. December is the last month of the year. _____

   c. The second month of the year is June. _____

   d. October comes after November. _____

   e. February comes before March. _____

   f. Christmas is in December. _____

3. Draw something you do in:

   a. December (winter)       b. June (summer)

97

| **Name** | **Date** |
|---|---|

1. Name the four seasons.

   _____            _____

   _____            _____

2. Match these.

   | June      July |
   |:---:|
   | August |

   Spring

   Winter

   | December      January |
   |:---:|
   | February |

   Summer

   | March      April |
   |:---:|
   | May |

   Autumn

   | September      October |
   |:---:|
   | November |

3. Complete each sentence using *winter, summer, spring,* or *fall.*

   **a.** The hottest season is            _____ .

   **b.** The coldest season is            _____ .

   **c.** Christmas is in            _____ .

   **d.** Your birthday is in            _____ .

   **e.** The season at the moment is            _____ .

   **f.** The next season is            _____ .

   **g.** After winter comes            _____ .

   **h.** Summer comes after            _____ .

| **Name** | **Date** |
|---|---|

1. Draw two things that take about one hour to do.

2. Count how many times you can do these activities in one minute.

    **a.** bounce a ball _____     **b.** write your name _____

    **c.** jump on the spot _____     **d.** clap _____

3. Color the things that take only seconds to do.

*#8988 Targeting Math: Measurement*

| **Name** | **Date** |
|---|---|

1. Complete the sentences using *longer* or *shorter*.

   **a.** It takes a _____ time to brush your teeth than to read a book.

   **b.** It takes a _____ time for a flower to grow than to clap.

2. Write the days of the week in order.

   **a.** _____   **d.** _____   **g.** _____

   **b.** _____   **e.** _____

   **c.** _____   **f.** _____

3. Write the months of the year in order.

   **a.** _____   **e.** _____   **i.** _____

   **b.** _____   **f.** _____   **j.** _____

   **c.** _____   **g.** _____   **k.** _____

   **d.** _____   **h.** _____   **l.** _____

4. Use the months above to:

   **a.** Circle the summer months in red.   **b.** Circle the winter months in blue.

   **c.** Circle the spring months in green.   **d.** Circle the autumn months in brown.

5. Write something that takes:

   **a.** about one hour to do.          **b.** about one minute to do.

   _____          _____

# MORE TIME

## Unit 2

*Calendar*
*Analog*
*Digital*
*Half past*

## Objectives

- *locates dates and events on a calendar*
- *recognizes and reads o'clock on analog clocks*
- *understands that clocks are used to tell the time of day*
- *tells the time on the hour using digital and/or analog clocks*
- *recognizes that daily activities are related to and influenced by time as indicated by clocks*
- *recognizes half past times on digital and analog clocks*
- *tells the time on a digital and analog clock in hours and minutes*

## Language

*after, analog, before, calendar, dates, day, digital, first, half past, hour, how many, lasts, month, o'clock, past, quarter to, quarter past, time, weekday, weekend*

## Materials/Resources

*colored pencils, digital and analog clocks, watches*

## Contents of Student Pages

* *Materials needed for each reproducible student page*

### Remember

❑ *Relate classroom activities to passing of time whenever possible.*

❑ *Make students aware that watches are types of clocks.*

❑ *Use a variety of words when discussing time.*

## Additional Activities

❑ List different places where you would find digital and analog clocks.

❑ Investigate different informal units used to measure the passing of time (e.g., egg timer, sundial, water clock, pendulum).

❑ Make a digital clock. In pairs, practice telling the time to each other.

❑ Find pictures of different kinds of clocks in magazines, brochures, and newspapers. Make a clock collage.

❑ Play Concentration/Memory or Go Fish. Write time in words on one set of blank cards and analog and digital clocks on other.

❑ Play Tic-Tac-Toe with two teams. Ask teams questions about time. If the answer is correct, they can put either a circle or cross on the grid. The first team to get three in a row is the winner.

## Answers

### Page 103  Calendars

1. a. Tuesday
   b. August 1
   c. Thursday
   d. Make sure the 5, 6, 12, 13, 19, 20, 26, and the 27 are colored red.
   e. Make sure all other days are colored blue.
   f. 23
   g. 31
   h. 5
2. Teacher check to see if calendar is filled in correctly.
   a. June          c. May
   b. July          d. 30

### Page 104  Analog Time

1. a. 7 o'clock      d. 11 o'clock
   b. 5 o'clock      e. 2 o'clock
   c. 6 o'clock      f. 12 o'clock

2.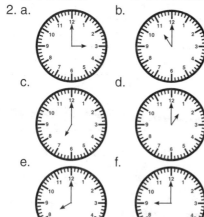

### Page 105  Digital Time

1. a. 11 o'clock   c. 5 o'clock   e. 9 o'clock
   b. 2 o'clock    d. 6 o'clock   f. 4 o'clock
2. a. 7:00   c. 3:00   e. 8:00
   b. 10:00   d. 12:00   f. 9:00

### Page 106  Digital and Analog

1. a. Clock face should show 9 o'clock, Digital clock should show 9:00.
   b. Clock face should show 12 o'clock, Digital clock should show 12:00.
   c. Clock face should show 7 o'clock, Digital clock should show 7:00.
   d. Clock face should show 11 o'clock, Digital clock should show 11:00.
2. a. Clock face should show 6 o'clock, Digital clock should show 6:00.
   b. 3 should be filled in the sentence, Digital clock should show 3:00.
   c. 4 should be filled in the sentence, Clock face should show 4 o'clock.
   d. 10 should be filled in the sentence, Digital clock should show 10:00.

### Page 107  Half Past

1.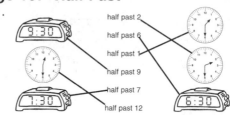

2. a. half past 5, 5:30, clock face should show 5:30.
   b. half past 6, 6:30, clock face should show 6:30.
   c. half past 12, 12:30, clock face should show 12:30.
   d. half past 1, 1:30, clock face should show 1:30.
   e. half past 10, 10:30, clock face should show 10:30.

### Page 108  5-Minute Intervals

1. a. 9:05, Make sure clock face matches.
   b. 6:20, Make sure clock face matches.
   c. 11:15, Make sure clock face matches.
   d. 11:50, Make sure clock face matches.
   e. 5:45, Make sure clock face matches.
   f. 12:35, Make sure clock face matches.

2.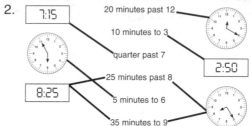

### Page 109  Assessment

1. a. Sunday        c. Tuesday
   b. January 1     d. 5

2.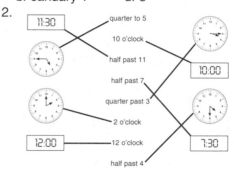

### Page 110  Activity—Reading Times

Hans = 6 hours, Jenny = 1 hour, Lin = 3 hours, Pablo = 4 hours

| Name | Date |
|------|------|

1. Below is a calendar for August.

| Sunday | Monday | Tuesday | Wednesday | Thursday | Friday | Saturday |
|--------|--------|---------|-----------|----------|--------|----------|
|        |        | 1 | 2 | 3 | 4 | 5 |
| 6 | 7 | 8 | 9 | 10 | 11 | 12 |
| 13 | 14 | 15 | 16 | 17 | 18 | 19 |
| 20 | 21 | 22 | 23 | 24 | 25 | 26 |
| 27 | 28 | 29 | 30 | 31 |  |  |

   **a.** Which is the first day of the month? _____

   **b.** What is the date of the first day of the month? _____

   **c.** Which day is the last day of the month? _____

   **d.** Color the weekends red.      **e.** Color the weekdays blue.

   **f.** Count the weekdays._____      **g.** How many days in August?_____

   **h.** How many Wednesdays are in August? _____

2. Fill in the calendar.

| Sunday | Monday | Tuesday | Wednesday | Thursday | Friday | Saturday |
|--------|--------|---------|-----------|----------|--------|----------|
|        |        | 2 |   | 4 |    | 6 |
|   | 8 |   |   |   | 12 |   |
|   | 15 |   |   |   |    | 20 |
| 21 |   |   | 24 |   |    |   |
|   | 29 | 30 |   |   |    |   |

   **a.** This is the 6th month.  What is its name? _____

   **b.** Which month comes after this one? _____

   **c.** Which month comes before this one? _____

   **d.** How many days in this month? _____

(103)

| **Name** | **Date** |
|---|---|

**1.** What time is it?

**a.**

_____ o'clock

**b.**

_____ o'clock

**c.**

_____ o'clock

**d.**

_____ o'clock

**e.**

_____ o'clock

**f.**

_____ o'clock

**2.** Draw hands on these clocks to show the times.

**a.**

3 o'clock

**b.**

11 o'clock

**c.**

7 o'clock

**d.**

1 o'clock

**e.**

8 o'clock

**f.**

9 o'clock

**104**

| **Name** | **Date** |
| --- | --- |

**1.** What time is it?

**a.**

\_\_\_\_\_ o'clock

**b.**

\_\_\_\_\_ o'clock

**c.**

\_\_\_\_\_ o'clock

**d.**

\_\_\_\_\_ o'clock

**e.**

\_\_\_\_\_ o'clock

**f.**

\_\_\_\_\_ o'clock

**2.** Fill in the times on the digital clocks.

**a.**

7 o'clock

**b.**

10 o'clock

**c.**

3 o'clock

**d.**

12 o'clock

**e.**

8 o'clock

**f.**

9 o'clock

105

| **Name** | **Date** |
|---|---|

**1.** Complete.

> **Example:** I go to bed at 8 o'clock.
>
>

**a.** School starts at 9 o'clock.

**b.** Midnight is at 12 o'clock.

**c.** I wake up at 7 o'clock.

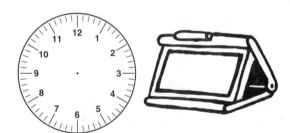

**d.** I eat lunch at 11 o'clock.

**2.** Complete.

**a.** I eat dinner at ___6___ o'clock.

**b.** School ends at _____ o'clock.

**c.** We play tennis at _____ o'clock.

**d.** Soccer starts at _____ o'clock.

| Name | Date |
|------|------|

**1.** Match these.

half past 2

half past 6

half past 1

half past 9

half past 7

half past 12

**2.** Complete these.

**Example:** Half an hour before 3 o'clock is

_____ half past 2 _____

   2:30

**a.** Half an hour after 5 o'clock is

_____ .

**b.** Half an hour before 7 o'clock is

_____ .

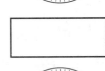

**c.** Half an hour after 12 o'clock is

_____ .

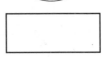

**d.** 30 minutes after 1 o'clock is

_____ .

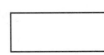

**e.** 30 minutes before 11 o'clock is

_____ .

| **Name** | **Date** |
|---|---|

1. Complete these.

10 minutes past 3          5 minutes to 4

 3:10           3:55

**a.** 5 minutes past 9     **b.** 20 minutes past 6     **c.** quarter past 11

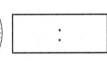 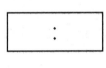      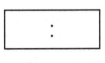

**d.** 10 minutes to 12     **e.** quarter to 6     **f.** 25 minutes to 1

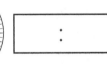       

2. Match these.

7:15

8:25

20 minutes past 12

10 minutes to 3

quarter past 7

25 minutes past 8

5 minutes to 6

35 minutes to 9

2:50

| Name | Date |
|------|------|

1. Below is a calendar for January.

| Sunday | Monday | Tuesday | Wednesday | Thursday | Friday | Saturday |
|--------|--------|---------|-----------|----------|--------|----------|
| 1 | 2 | 3 | 4 | 5 | 6 | 7 |
| 8 | 9 | 10 | 11 | 12 | 13 | 14 |
| 15 | 16 | 17 | 18 | 19 | 20 | 21 |
| 22 | 23 | 24 | 25 | 26 | 27 | 28 |
| 29 | 30 | 31 | | | | |

    **a.** Which day is the first day of the month? _____

    **b.** What is the date of the first day of the month? _____

    **c.** What is the last day of the month? _____

    **d.** How many Mondays in January? _____

2. Match these.

quarter to 5

10 o'clock

half past 11

half past 7

quarter past 3

2 o'clock

12 o'clock

half past 4

*#8988 Targeting Math: Measurement*

| **Name** | **Date** |
|---|---|

Four children read different books. It took each one a different time to finish reading their book—1 hour, 3 hours, 4 hours, and 6 hours.

Use these clues to work out how long each person took to read their book.

1. Lin took longer than Jenny but less time than Hans.

2. Pablo finished reading before Hans.

3. Jenny was finished first.

4. Pablo was still reading when Lin had finished.

Write how long it took each person to read his or her book.

Hans _____        Jenny _____

Lin _____        Pablo _____

# Skills Index

The following index lists specific objectives for the student pages of each unit in the book. The objectives are grouped according to the sections listed in the Table of Contents. Use the Skills Index as a resource for identifying the units and student pages you wish to use.

## Area

estimate, compare, and order the areas of shapes using informal units (Pages 8, 13, 18, 19)

estimate, compare, and order the areas of shapes using formal units (Page 22)

estimate the order of things by length, area, mass, and capacity and make numerical estimates of length using a unit that can be seen or handled (Pages 20, 21)

explain simple mathematical situations using everyday language, actions, materials, and drawing (Page 13)

make non-numerical estimates of size involving everyday movements and actions (Pages 9, 10, 11, 12)

recognize and compare the sizes of groups through a variety of strategies such as estimating, matching one-to-one, and counting (Page 8)

respond to and use everyday comparative language relating to area when recording and communicating measurements (Pages 9, 12)

sort and describe objects in terms of their features such as size and shape (Page 11)

## Capacity/Volume

approximate, count, compare, order, and represent whole numbers and groups of objects up to 100 (Page 38)

ask and respond to mathematical questions drawing, describing, acting, telling, guessing and checking, and retelling (Page 29)

count collections up to 10 and beyond (Page 40)

describe and model the relationships between the parts and the whole (Pages 32, 42)

estimate, compare, and order the capacity of containers using informal units (Pages 31, 32, 38, 39, 42)

order whole numbers up to 999 (Page 37)

read and interpret graphs made from objects (Page 41)

represent addition and subtraction facts up to 20 using concrete materials and in symbolic form (Page 29)

sort and describe objects in terms of their features such as size and shape (Pages 27, 28, 30, 40)

use available technology to help in the solution of mathematical problems (Page 37)

use number skills involving whole numbers to solve problems (Page 37)

## Length

chooses the appropriate physical attribute when comparing and measuring things and units which relate to that attribute (Page 67)

compares and orders length using indirect methods (Page 59)

counts collections up to ten objects and beyond (Page 58)

describes relative distances using such terms as "nearer" and "further" (Page 55)

directly and indirectly compares lengths and measures and makes lengths and distances by counting uniform units including centimeters (Page 66)

estimates, compares, orders, and measures the length of objects and the distance between them using informal or formal units (Pages 58, 60)

explores 2-D shapes and 3-D objects, describing them in comparative language (Page 47)

finds things bigger or smaller than a given object (Pages 48, 49, 51)

measures perimeter (Page 67)

recognizes what features of an object can be measured (Page 65)

recognizes a yard length (Page 68)

represents, interprets, and explains mathematical technology, including simple graphs and diagrams (Page 69)

responds to and uses everyday comparative language relating to length when recording and communicating measurements (Pages 50, 55)

sorts and describes objects in terms of their features, such as size and shape (Pages 47, 48, 49, 56, 57)

*111*

# Skills Index

tests estimates of physical quantities using appropriate informal or formal units (Pages 58, 65, 68)

understands everyday comparative language associated with length (Pages 49, 50, 57)

uses appropriate language for each length, capacity, and mass (Page 51)

uses formal units (Page 67)

uses guessing and checking for length comparisons of two objects (Page 60)

uses formal units carefully and consistently to measure and compare length (Page 59)

uses matching to identify longer than, shorter than, and same length objects (Page 56)

## Mass

answers mathematical questions using objects, pictures, imagery, actions, or trial and error (Page 80)

comments on information in displays produced by himself/herself or others (Pages 85, 86)

counts collections up to 10 objects and beyond (Page 87)

demonstrates an understanding that numbers can be represented using groups of 10, 100, and 1,000 (Page 84)

describes and models the relationships between the parts and the whole (Page 88)

engages in writing texts with the intention of conveying an idea or message (Page 78)

estimates, compares, orders, and measures the mass of objects using informal units
(Pages 75, 77, 78, 79, 80, 84, 85, 86, 87)

makes block graphs using one-to-one correspondence between real data and a representation (Page 86)

makes judgments of comparison of mass by lifting (Page 87)

models numbers and number relationships in a variety of ways, and uses them in solving number problems (Page 88)

orders whole numbers to 999 (Page 84)

sorts and describes objects in terms of their features such as size and shape (Pages 75, 76)

uses number skills involving whole numbers to solve problems (Pages 79, 88)

## Time

counts informal units to measure a period of time (Page 99)

locates dates and events on a calendar (Page 103)

orders days of the week or months of the year (Page 97)

recognizes that daily activities are related to and influenced by time as indicated by seasons (Page 98)

relates days of the week or months of the year to events in his/her own life (Pages 96, 97)

recognizes and reads o'clock on analog clocks (Page 104)

recognizes half past times on digital and analog clocks (Page 107)

recognizes that daily activities are related to and influenced by time as indicated by clocks (Page 106)

responds to or uses everyday language associated with time (Pages 95, 97, 98)

spells correctly and orders days of the week (Page 96)

tells the time on the hour using digital and/or analog clocks (Pages 105, 106)

tells the time on a digital and analog clock in hours and minutes (Page 108)

understands that clocks are used to reflect the passage of time (Page 99)

understands that clocks are used to tell the time of the day (Pages 104, 105, 106)

uses everyday language to compare the duration of events (Page 94)

#8988 Targeting Math: Measurement